# animal sciences

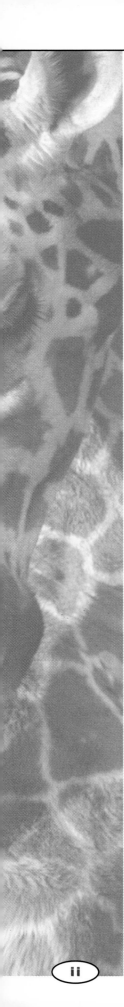

# animal sciences

VOLUME **4**
**Per–Zoo**

**Allan B. Cobb, Editor in Chief**

**MACMILLAN REFERENCE USA**

**GALE GROUP**

**THOMSON LEARNING**

*New York • Detroit • San Diego • San Francisco*
*Boston • New Haven, Conn. • Waterville, Maine*
*London • Munich*

Macmillan Reference USA          The Gale Group
1633 Broadway                    27500 Drake Rd.
New York, NY 10019               Farmington Hills, 48331-3535

Printed in the United States of America
1 2 3 4 5 6 7 8 9 10

**Library of Congress Cataloging-in-Publication Data**
Animal Sciences / Allan B. Cobb, editor in chief.
        p.   cm.
Includes bibliographical references and index.
        ISBN: 0-02-865556-7 (set) - ISBN 0-02-865557-5 (vol. 1) - ISBN
0-02-865558-3 (vol. 2) - ISBN 0-02-865559-1 (vol. 3) - ISBN 0-02-865560-5
(vol. 4)
1. Animal Culture. 2. Livestock. I. Title.
SF61.C585 2001
590.3-dc21                                          2001026627
                                                          Rev

## Geological Time Scale

**Time Range of Several Groups of Plants & Animals**

(bar chart)
- Birds
- Mammals
- Reptiles
- Amphibians
- Land Plants
- Fishes
- Invertebrates

| Eon | Era | Period | Epoch | Million Years Before Present | Significant Events |
|---|---|---|---|---|---|
| Phanerozoic | Cenozoic | Quaternary | Holocene | 0.01 | recorded human history, rise and fall of civilizations, global warming, habitat destruction, pollution mass extinction |
| | | | Pleistocene | 1.6 | *Homo sapiens*, ice ages |
| | | Tertiary | Pliocene | 5.3 | global cooling, savannahs, grazing mammals |
| | | | Miocene | 24 | global warming, grasslands, *Chalicotherium* |
| | | | Oligocene | 37 | |
| | | | Eocene | 58 | modern mammals flourish, ungulates |
| | | | Paleocene | 66 | |
| | Mesozoic | Cretaceous | | 144 | last of age of dinosaurs, modern mammals appear, flowering plants, insects |
| | | Jurassic | | 208 | huge plant-eating dinosaurs, carnivorous dinosaurs, first birds, breakup of Pangea |
| | | Triassic | | 245 | lycophytes, glossopterids, and dicynodonts, and the dinosaurs |
| | Paleozoic | Permian | | 286 | Permian ends with largest mass extinction in history of Earth, most marine invertebrates extinct |
| | | Pennsylvanian | | 320 | vast coal swamps, evolution of amniote egg allowing exploitation of land |
| | | Missipian | | 360 | shallow seas cover most of Earth |
| | | Devonian | | 408 | vascular plants, the first tetrapods, wingless insects, arachnids, brachiopods, corals, and ammonite were also common, many new kinds of fish appeared |
| | | Silurian | | 438 | Coral reefs, rapid spread of jawless fish, first freshwater fish, first fish with jaws, first good evidence of life on land, including relatives of spiders and centipedes |
| | | Ordovician | | 505 | most dry land collected into Gondwana, many marine invertebrates, including graptolites, trilobites, brachiopods, and the conodonts (early vertebrates), red and green algae, primitive fish, cephalopods, corals, crinoids, and gastropods, possibly first land plants |
| | | Cambrian | | 570 | most major groups of animals first appear, Cambrian explosion |
| Precambrian | Proterozoic | | | 2500 | stable continents first appear, first abundant fossils of living organisms, mostly bacteria and archeobacteria, first eukaryotes, first evidence of oxygen build-up |
| | Archean | | | 3800 | atmosphere of methane, ammonia, rocks and continental plates began to form, oldest fossils consist of bacteria microfossils stromatolites, colonies of photosynthetic bacteria |
| | Hadean | | | 4500 | pre-geologic time, Earth in formation |

# COMPARISON OF THE FIVE-KINGDOM AND SIX-KINGDOM CLASSIFICATION OF ORGANISMS

| Five Kingdom | Six Kingdom |
|---|---|

Kingdom: Monera                                                    Kingdom: Archaebacteria
    Phylum: Bacteria                                           Kingdom: Eubacteria
    Phylum: Blue-green algae (cyanobacteria)

Kingdom: Protista
    Phylum: Protozoans
        Class: Ciliophora
        Class: Mastigophora
        Class: Sarcodina
        Class: Sporozoa
    Phylum: Euglenas
    Phylum: Golden algae and diatoms
    Phylum: Fire or golden brown algae
    Phylum: Green algae
    Phylum: Brown algae
    Phylum: Red algae
    Phylum: Slime molds
Kingdom: Fungi
    Phylum: Zygomycetes
    Phylum: Ascomycetes
    Phylum: Basidiomycetes
Kingdom: Plants
    Phylum: Mosses and liverworms
    Phylum: Club mosses
    Phylum: Horsetails
    Phylum: Ferns
    Phylum: Conifers
    Phylum: Cone-bearing desert plants
    Phylum: Cycads
    Phylum: Ginko
    Phylum: Flowering plants
        Subphylum: Dicots (two seed leaves)
        Subphylum: Monocots (single seed leaves)
Kingdom: Animals
    Phylum: Porifera
    Phylum: Cnidaria
    Phylum: Platyhelminthes
    Phylum: Nematodes
    Phylum: Rotifers
    Phylum: Bryozoa
    Phylum: Brachiopods
    Phylum: Phoronida
    Phylum: Annelids
    Phylum: Mollusks
        Class: Chitons
        Class: Bivalves
        Class: Scaphopoda
        Class: Gastropods
        Class: Cephalopods
    Phylum: Arthropods
        Class: Horseshoe crabs
        Class: Crustaceans
        Class: Arachnids
        Class: Insects
        Class: Millipedes and centipedes
    Phylum: Echinoderms
    Phylum: Hemichordata
    Phylum: Cordates
        Subphylum: Tunicates
        Subphylum: Lancelets
        Subphylum: Vertebrates
            Class: Agnatha (lampreys)
            Class: Sharks and rays
            Class: Bony fishes
            Class: Amphibians
            Class: Reptiles
            Class: Birds
            Class: Mammals
                Order: Monotremes
                Order: Marsupials
            Subclass: Placentals
                Order: Insectivores
                Order: Flying lemurs
                Order: Bats
                Order: Primates (including humans)
                Order: Edentates
                Order: Pangolins
                Order: Lagomorphs
                Order: Rodents
                Order: Cetaceans
                Order: Carnivores
                Order: Seals and walruses
                Order: Aardvark
                Order: Elephants
                Order: Hyraxes
                Order: Sirenians
                Order: Odd-toed ungulates
                Order: Even-toed ungulates

# PHYLOGENETIC TREE OF LIFE

This diagram represents the phylogenetic relationship of living organisms, and is sometimes called a "tree of life." Often, these diagrams are drawn as a traditional "tree" with "branches" that represent significant changes in the development of a line of organisms. This phylogenetic tree, however, is arranged in a circle to conserve space. The center of the circle represents the earliest form of life. The fewer the branches between the organism's name and the center of the diagram indicate that it is a "lower" or "simpler" organism. Likewise, an organism with more branches between its name and the center of the diagram indicates a "higher" or "more complex" organism. All of the organism names are written on the outside of the circle to reinforce the idea that all organisms are highly evolved forms of life.

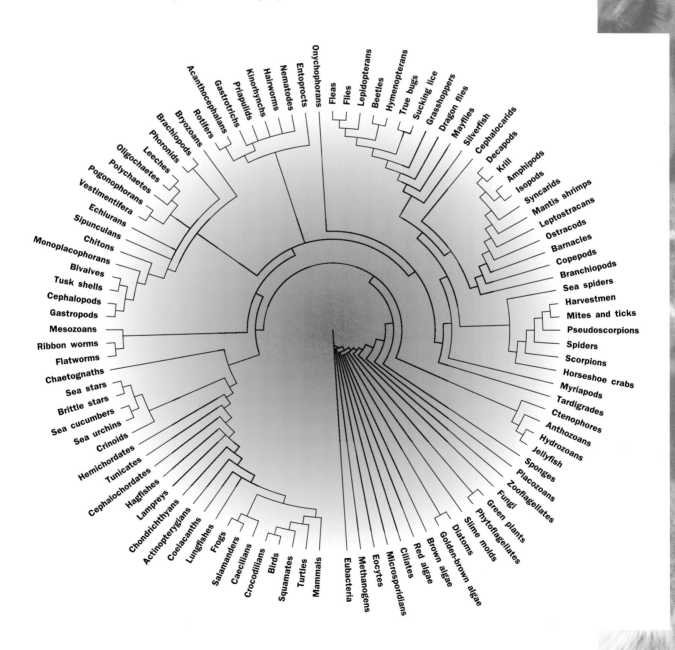

## SI BASE AND SUPPLEMENTARY UNIT NAMES AND SYMBOLS

| Physical Quality | Name | Symbol |
|---|---|---|
| Length | meter | m |
| Mass | kilogram | kg |
| Time | second | s |
| Electric current | ampere | A |
| Thermodynamic temperature | kelvin | K |
| Amount of substance | mole | mol |
| Luminous intensity | candela | cd |
| Plane angle | radian | rad |
| Solid angle | steradian | sr |

### Temperature

Scientists commonly use the Celsius system. Although not recommended for scientific and technical use, earth scientists also use the familiar Fahrenheit temperature scale (ºF). 1ºF = 1.8ºC or K. The triple point of $H_2O$, where gas, liquid, and solid water coexist, is 32ºF.

- To change from Fahrenheit (F) to Celsius (C):
  ºC = (ºF-32)/(1.8)
- To change from Celsius (C) to Fahrenheit (F):
  ºF = (ºC x 1.8) + 32
- To change from Celsius (C) to Kelvin (K):
  K = ºC + 273.15
- To change from Fahrenheit (F) to Kelvin (K):
  K = (ºF-32)/(1.8) + 273.15

## UNITS DERIVED FROM SI, WITH SPECIAL NAMES AND SYMBOLS

| Derived Quantity | Name of SI Unit | Symbol for SI Unit | Expression in Terms of SI Base Units |
|---|---|---|---|
| Frequency | hertz | Hz | $s^{-1}$ |
| Force | newton | N | $m\ kg\ s^{-2}$ |
| Pressure, stress | Pascal | Pa | $N\ m^{-2}$ $=m^{-1}\ kg\ s^{-2}$ |
| Energy, work, heat | Joule | J | $N\ m$ $=m^2\ kg\ s^{-2}$ |
| Power, radiant flux | watt | W | $J\ s^{-1}$ $=m^2\ kg\ s^{-3}$ |
| Electric charge | coulomb | C | $A\ s$ |
| Electric potential, electromotive force | volt | V | $J\ C^{-1}$ $=m^{-2}\ kg\ s^{-3}\ A^{-1}$ |
| Electric resistance | ohm | _ | $V\ A^{-1}$ $=m^2\ kg\ s^{-3}\ A^{-2}$ |
| Celsius temperature | degree Celsius | C | K |
| Luminous flux | lumen | lm | cd sr |
| Illuminance | lux | lx | $cd\ sr\ m^{-2}$ |

## UNITS USED WITH SI, WITH NAME, SYMBOL, AND VALUES IN SI UNITS

The following units, not part of the SI, will continue to be used in appropriate contexts (e.g., angtsrom):

| Physical Quantity | Name of Unit | Symbol for Unit | Value in SI Units |
|---|---|---|---|
| Time | minute | min | 60 s |
| | hour | h | 3,600 s |
| | day | d | 86,400 s |
| Plane angle | degree | ° | $(\pi/180)$ rad |
| | minute | ' | $(\pi/10,800)$ rad |
| | second | " | $(\pi/648,000)$ rad |
| Length | angstrom | Å | $10^{-10}$ m |
| Volume | liter | l, L | $1\ dm^3 = 10^{-3}\ m^3$ |
| Mass | ton | t | $1\ mg = 10^3\ kg$ |
| | unified atomic mass unit | $u\ (=m_a(^{12}C)/12)$ | $\approx 1.66054 \times 10^{-27}$ kg |
| Pressure | bar | bar | $10^5\ Pa = 10^5\ N\ m^{-2}$ |
| Energy | electronvolt | $eV\ (= e \times V)$ | $\approx 1.60218 \times 10^{-19}$ J |

## CONVERSIONS FOR STANDARD, DERIVED, AND CUSTOMARY MEASUREMENTS

### Length

| | |
|---|---|
| 1 angstrom (Å) | 0.1 nanometer (exactly) |
| | 0.000000004 inch |
| 1 centimeter (cm) | 0.3937 inches |
| 1 foot (ft) | 0.3048 meter (exactly) |
| 1 inch (in) | 2.54 centimeters (exactly) |
| 1 kilometer (km) | 0.621 mile |
| 1 meter (m) | 39.37 inches |
| | 1.094 yards |
| 1 mile (mi) | 5,280 feet (exactly) |
| | 1.609 kilometers |
| 1 astronomical unit (AU) | $1.495979 \times 10^{13}$ cm |
| 1 parsec (pc) | 206,264.806 AU |
| | $3.085678 \times 10^{18}$ cm |
| | 3.261633 light-years |
| 1 light-year | $9.460530 \times 10^{17}$ cm |

### Area

| | |
|---|---|
| 1 acre | 43,560 square feet (exactly) |
| | 0.405 hectare |
| 1 hectare | 2.471 acres |
| 1 square centimeter (cm²) | 0.155 square inch |
| 1 square foot (ft²) | 929.030 square centimeters |
| 1 square inch (in²) | 6.4516 square centimeters (exactly) |
| 1 square kilometer (km²) | 247.104 acres |
| | 0.386 square mile |
| 1 square meter (m²) | 1.196 square yards |
| | 10.764 square feet |
| 1 square mile (mi²) | 258.999 hectares |

## MEASUREMENTS AND ABBREVIATIONS

### Volume

| | |
|---|---|
| 1 barrel (bbl)*, liquid | 31 to 42 gallons |
| 1 cubic centimeter (cm³) | 0.061 cubic inch |
| 1 cubic foot (ft³) | 7.481 gallons |
| | 28.316 cubic decimeters |
| 1 cubic inch (in³) | 0.554 fluid ounce |
| 1 dram, fluid (or liquid) | ⅛ fluid ounce (exactly) |
| | 0.226 cubic inch |
| | 3.697 milliliters |
| 1 gallon (gal) (U.S.) | 231 cubic inches (exactly) |
| | 3.785 liters |
| | 128 U.S. fluid ounces (exactly) |
| 1 gallon (gal) (British Imperial) | 277.42 cubic inches |
| | 1.201 U.S. gallons |
| | 4.546 liters |
| 1 liter | 1 cubic decimeter (exactly) |
| | 1.057 liquid quarts |
| | 0.908 dry quart |
| | 61.025 cubic inches |
| 1 ounce, fluid (or liquid) | 1.805 cubic inches |
| | 29.573 milliliters |
| 1 ounce, fluid (fl oz) (British) | 0.961 U.S. fluid ounce |
| | 1.734 cubic inches |
| | 28.412 milliliters |
| 1 quart (qt), dry (U.S.) | 67.201 cubic inches |
| | 1.101 liters |
| 1 quart (qt), liquid (U.S.) | 57.75 cubic inches (exactly) |
| | 0.946 liter |

### Units of mass

| | |
|---|---|
| 1 carat (ct) | 200 milligrams (exactly) |
| | 3.086 grains |
| 1 grain | 64.79891 milligrams (exactly) |
| 1 gram (g) | 15.432 grains |
| | 0.035 ounce |
| 1 kilogram (kg) | 2.205 pounds |
| 1 microgram (μg) | 0.000001 gram (exactly) |
| 1 milligram (mg) | 0.015 grain |
| 1 ounce (oz) | 437.5 grains (exactly) |
| | 28.350 grams |
| 1 pound (lb) | 7,000 grains (exactly) |
| | 453.59237 grams (exactly) |
| 1 ton, gross or long | 2,240 pounds (exactly) |
| | 1.12 net tons (exactly) |
| | 1.016 metric tons |
| 1 ton, metric (t) | 2,204.623 pounds |
| | 0.984 gross ton |
| | 1.102 net tons |
| 1 ton, net or short | 2,000 pounds (exactly) |
| | 0.893 gross ton |
| | 0.907 metric ton |

### Pressure

| | |
|---|---|
| 1 kilogram/square centimeter (kg/cm²) | 0.96784 atmosphere (atm) |
| | 14.2233 pounds/square inch (lb/in²) |
| | 0.98067 bar |
| 1 bar | 0.98692 atmosphere (atm) |
| | 1.02 kilograms/square centimeter (kg/cm²) |

* There are a variety of "barrels" established by law or usage. For example, U.S. federal taxes on fermented liquors are based on a barrel of 31 gallons (141 liters); many state laws fix the "barrel for liquids" as 31½ gallons (119.2 liters); one state fixes a 36-gallon (160.5 liters) barrel for cistern measurment; federal law recognizes a 40-gallon (178 liters) barrel for "proof spirts"; by custom, 42 gallons (159 liters) comprise a barrel of crude oil or petroleum products for statistical purposes, and this equivalent is recognized "for liquids" by four states.

# Table of Contents

# Table of Contents

# animal sciences

# Permian

The Permian period, 280 to 230 million years ago, was named for the Perm Province of the Ural Mountains in Russia. The Permian signaled the end of the "ancient life" Paleozoic era.

In the Permian, the close ties between geology and evolution were especially apparent. The two great land masses of the Paleozoic drifted close enough together to form one supercontinent, Pangaea. Collisions in the tectonic plates created extensive volcanic activity and heaved up the Urals, Alps, Appalachians, and Rocky Mountains. The shallow inland seas drained to leave deposits of gypsum and salt. Vast sand dunes throughout much of what is now North America and Europe were recorded by massive yellow sandstones (hardened sand dunes) that contained few fossils other than scorpions.

Great glaciers scoured the southern regions of Africa, India, and Australia, further inhibiting life. **Conifers** and a few cold-hardy plants grew along the fringes of the immense ice cap.

The long stable **climate** of the Carboniferous gave way to dryness, with severe fluctuations of heat and cold. Only in the tropics of Pangaea did anything remain of the great Carboniferous rain forests, and there insects and amphibians continued to evolve.

Insects, members of the **arthropod** or "jointed leg" animals whose ancestors were the first to explore both land and air, continued to flourish in every new ecological opportunity. Several new groups appeared—the bugs, cicadas, and beetles. Thanks possibly to their diminutive size and adaptable **metamorphosis,** in which young live and feed in a totally different

**coniferous** having pine trees and other conifers

**climate** long-term weather patterns for a particular region

**arthropod** a phylum of invertebrates characterized by segmented bodies and jointed appendages such as antennae and legs

**metamorphosis** a drastic change from a larva to an adult

| Era | Period | Epoch | Million Years Before Present |
|-----|--------|-------|------------------------------|
| Paleozoic | Permian | | 286 |
| | Pennsylvanian | | 320 |
| | Missipian | | 360 |
| | Devonian | | 408 |
| | Silurian | | 438 |
| | Ordovician | | 505 |
| | Cambrian | | 570 |

The permian period and surrounding time periods.

An amphibian fossil from the Permian era on display at the Field Museum in Chicago, Illinois.

**ammonites** an extinct group of cephalopods with a curled shell

**brachiopods** a phylum of marine bivalve mollusks

**trilobites** an extinct class of arthropods

**herbivores** animals who eat plants only

**insectivores** animals who eats insects

**habitats** physical locations where an organism lives in an ecosystem

**genes** segments of DNA located on chromosomes that direct protein production

environment from adults, the arthropods became the most evolutionarily successful animals on Earth. Amphibians fared less well, mostly just hanging on in those areas still hospitable to their warm, moist requirements.

Many marine species thrived in the shallow seas. Thousands of types of sponges, corals, **ammonites**, bryozoans, **brachiopods**, and snails left their remains in the rocks that now make up the mountains of west Texas and southern New Mexico. Bony fishes remained plentiful. However, spiny fishes, the fleshy-finned rhipidistians (organisms who originally gave rise to amphibians), and the once-dominant **trilobites** disappeared.

Reptiles flourished in the semidesert regions that made up much of Pangaea. Their leathery-skinned, cold-blooded bodies were ideal for the hotter, drier climate. Reptile adaptations led to **herbivores** and **insectivores** who could exploit new food resources. As their legs continued to become stronger and more upright, the reptiles increased in body size and mobility. *Coelorosauravus* joined the flying insects, gliding from tree to tree by means of a sail-like membrane. And *Mesosaurus*, a 1 meter (3 feet) long fish eater, returned to living underwater. Virtually the whole of Pangaea was dominated by the reptiles.

However, all this exuberance ended. The close of the Permian was marked by the worst extinction ever recorded. More than 75 percent of all plant and animal groups disappeared forever from the land, and in the ocean only about 5 percent of existing species survived. As devastating as these losses were, evolution and extinction are a recurring theme: the emptying of **habitats**, the reshuffling of **genes**, and a new start. Survival of the fittest might really be said to be survival of the luckiest. SEE ALSO GEOLOGICAL TIME SCALE.

*Nancy Weaver*

**Bibliography**

Asimov, Isaac. *Life and Time.* Garden City, NY: Doubleday & Company, 1978.

Fortey, Richard. *Fossils: The Key to the Past.* Cambridge, MA: Harvard University Press, 1991.

———. *Life: A Natural History of the First Four Billion Years of Life on Earth.* New York: Viking Press, 1998.

Friday, Adrian, and David S. Ingram, eds. *The Cambridge Encyclopedia of Life Sciences.* London: Cambridge University, 1985.

Gould, Stephen Jay, ed. *The Book of Life.* New York: W. W. Norton & Company, 1993.

McLoughlan, John C. *Synapsida: A New Look Into the Origin of Mammals.* New York: Viking Press, 1980.

Steele, Rodney, and Anthony Harvey, eds. *The Encyclopedia of Prehistoric Life.* New York: McGraw Hill, 1979.

Wade, Nicholas, ed. *The Science Times Book of Fossils and Evolution.* New York: The Lyons Press, 1998.

# Pesticide

Pesticides are natural or human-made substances used to kill pest species such as rodents and insects. It is not surprising that many of these substances are highly toxic not only to the pests, but to other biological organisms as well. Pesticides are used in forests, agricultural regions, parks, residential areas, and within the home.

The bulk of **pesticide** use is related to agricultural pest control. In fact, pesticide application increased dramatically when intensive agricultural methods began to be used near the start of the twentieth century. Although pesticides clearly help to increase agricultural production, they also harm humans and other animal species. In addition, they contaminate the environment, often persisting in water, air, and soil for long periods of time. The World Health Organization reports over one million human pesticide poisonings every year, including twenty thousand that result in death.

These numbers do not include the slower and more subtle effects that exposure to pesticides can have on human health. Many pesticides, for example, are carcinogenic, or cancer-causing. Finally, because pest species always evolve resistance to pesticides over time, ever-increasing amounts or different types of pesticides are constantly required to maintain the same effect.

Some pesticides are inorganic, containing naturally toxic compounds such as lead, arsenic, or mercury. Because these chemicals cannot be broken down, they accumulate in the environment. Natural pesticides include substances produced by plants such as tobacco and certain conifer trees. These are used by the plant species that produce them to ward off herbivores. The majority of pesticides, however, are human-made organic chemicals that function by affecting some essential **physiological** function of pest species.

One of the best-known pesticides is dichloro-diphenyl-trichloroethane, commonly known as DDT. When DDT was first invented in 1939, by Swiss chemist Paul Muller, it was hailed as a major breakthrough in pesticide

**pesticide** any substance that controls the spread of harmful or destructive organisms

**physiological** describes the basic activities that occur in the cells and tissues of an animal

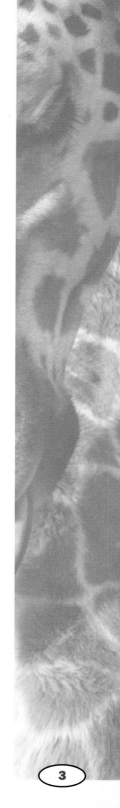

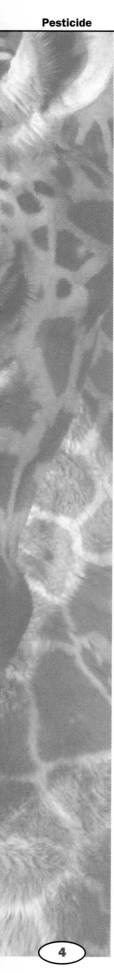

Pesticides are sprayed over farmland in Florida.

development. In fact, Muller received a Nobel Prize for the achievement. DDT found its first use in World War II, when it was sprayed in malarial areas to kill disease-carrying insects to safeguard U.S. troops.

After the war, DDT was widely used in the United States for agricultural control, and like many pesticides seemed highly effective at first. DDT was praised particularly for being highly toxic to insects while comparatively harmless for other species. DDT also had the advantages of being inexpensive to produce and easy to spray. By the 1950s, however, there was evidence that insect pests were evolving resistance to DDT. There were also hints that DDT might not be so harmless after all.

Rachel Carson's monumental book, *Silent Spring* (1962), was critical in bringing public attention to the serious side effects of DDT use for all living species. The title of the book refers to the absence of birdsong, a result of countless massive bird deaths throughout the country that Carson traced to DDT spraying. Studies of the impact of DDT have shown that the chemical breaks down very slowly, often lingering in the environment for decades after application. DDT is taken up by organisms through diet, and then accumulates in the fatty tissues. This effect is magnified higher up the food chain because any time a predator eats a prey item, the predator takes in all the DDT stored in the tissues of that prey, and then stores it in its own body.

**bio-accumulation** the build up of toxic chemicals in an organism

**endocrine systems** groupings of organs or glands that secrete hormones into the bloodstream

This process is called **bio-accumulation**. Bio-accumulation explains why birds high in the food chain, such as eagles, owls, and other birds of prey, are particularly vulnerable to DDT poisoning. DDT affects the **endocrine systems** of birds, throwing off the hormonal control of reproduction. Therefore, large amounts of bio-accumulated DDT cause the delay or cessation of egg laying. When eggs are produced, they are characterized by extremely thin eggshells that break easily during incubation. Although birds

appear to be particularly vulnerable to DDT, numerous other species are affected as well.

Carson also showed that there were causal links between pesticides, genetic **mutations**, and diseases such as cancer. Concerns regarding the tremendous health risks posed by DDT contributed to its being banned in the United States in 1972. Since then, many once-threatened species are now returning. *Silent Spring* is often credited not only with the ban of DDT, but with initiating awareness that toxic substances can be extremely harmful not only to the environment but to all the species that live within it, including humans. *Silent Spring* was crucial to the beginnings of environmentalism, as well as to the creation of the Environmental Protection Agency (EPA) in 1970.

Numerous pesticides are still in use now, including many that are even more toxic than DDT. Some of these break down more easily, however, and therefore do not remain in the environment for as long a period. Nonetheless, as awareness of some of the damaging cumulative effects of pesticides has increased, the popularity of and demand for organic foods has also increased.

In addition to toxicity, another problem with pesticide application is that pests inevitably evolve resistance. Pesticide resistance is a striking example of how efficiently **natural selection** can operate. In many cases, **alleles** that offer resistance to particular pesticides already exist in the **population** at very low frequencies. The application of pesticides selects strongly for these resistant alleles and causes them to spread quickly throughout the population. A classic example of the evolution of pesticide resistance is that of rats and warfarin. Warfarin is a pesticide that interferes with vitamin K and prevents blood coagulation, resulting in internal bleeding and death. Resistance to warfarin is conferred by a single gene, which spreads quickly through the rat population upon large-scale application of warfarin.

Because of the many harmful side effects of pesticide use, scientists have worked to develop alternative means for pest control. These include mechanical strategies such as screens or traps, the development of pest-resistant plants, crop cycling, and **biological control**, which aims to control pest populations by releasing large numbers of predators or parasites of a pest. In general, thorough information on the natural history of pest species, such as its life cycle requirements and natural enemies, helps to provide insight into the sort of strategies that may be effective in controlling it. SEE ALSO CARSON, RACHEL; DDT; SILENT SPRING.

*Jennifer Yeh*

**mutations** abrupt changes in the genes of an organism

**natural selection** the process by which organisms best suited to their environment are most likely to survive and reproduce

**alleles** two or more alternate forms of a gene

**population** a group of individuals of one species that live in the same geographic area

**biological control** the introduction of natural enemies such as parasites, predators, or pathogens as a method of controlling pests instead of using chemicals

**Bibliography**

Carson, Rachel. *Silent Spring*. Boston: Houghton Mifflin, 1962.

Gould, James L., and William T. Keeton. *Biological Science*, 6th ed. New York: W. W. Norton, 1996.

**Internet Resources**

*Office of Pesticide Programs*. United States Environment Protection Agency. <http://www.epa.gov/pesticides/>.

# Phylogenetic Relationships of Major Groups

**phylogeny** the evolutionary history of a species or group of related species

**systematic** study of the diversity of life

The evolutionary history of a species or group of related species is called its **phylogeny**. When the evolutionary history is diagrammed, it is shown in the shape of a tree that traces evolutionary relationships as they have changed over time. The reconstruction of phylogenetic history is part of the field of **systematics**. The diversity of the phylogenetic tree is a reflection of speciation.

A phylogenetic tree shows not only how closely related two groups are but also how once-related species evolved independently. The further back in time a group branched represents a greater amount of time for divergent evolution to occur. When systematists construct a phylogenetic tree, they

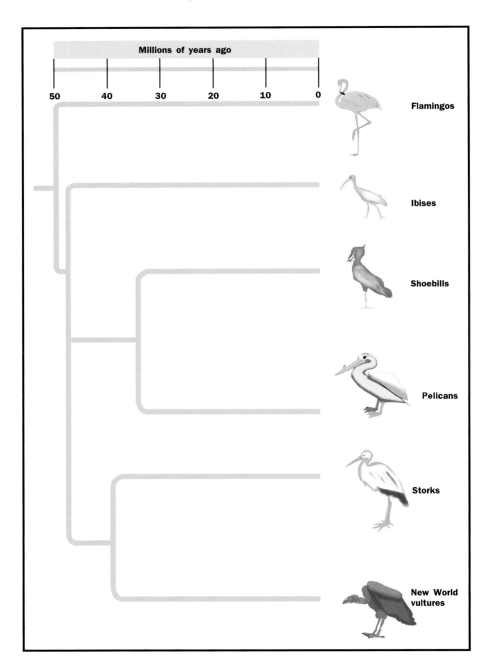

Study of the DNA sequences of these birds explains their evolutionary relationships.

consider as much data as possible. Whenever possible, they take the **fossil record** into account to identify when branching occurred. Scientists can compare ribosomal RNA or **mitochondrial DNA** of different organisms to pinpoint branches in the evolutionary history. After all the available data are compiled, the relationship can be drawn as a phylogenetic tree. Scientists often revise phylogenetic trees as new techniques or new data further clarify evolutionary relationships.

A phylogenetic tree can be used to show the evolutionary relationships of different groups. The figure (opposite) shows a simple phylogenetic tree of selected bird families. The first branch of the tree, which branched off about fifty million years ago, leads to the modern flamingo family on one branch and the other families off the other branch. The shoebill and the pelican families branched off about forty-five million years ago. It is important to realize that even though flamingos branched off much earlier than pelicans, each family has been influenced by evolutionary change. As environmental pressures such as climate change took place, each family adapted to the changes. This phylogenetic tree does not show any branches that became extinct.

Using a phylogenetic tree to find how closely related animals are is a relatively simple task. The closer together two **phyla** are on the tree, the more closely related the phyla. On the phylogenetic tree within the front-matter of this book, only the phyla are shown. Each phyla can be further divided into classes, orders, families, genera, and species. Also note that phyla that branch very close to the edge of the circle are closely related, while those that branch closer to the center of the tree are more distantly related. For example, frogs and **salamanders** are very closely related because they branch close to the outer edge of the circle. Eubacteria and methanogens are close together, but they branch close to the center of the tree. This means that they are distantly related.

*Allan B. Cobb*

### Bibliography

Barnes-Svarney, ed. *The New York Public Library Science Desk Reference*. New York: Macmillan, 1995.

Rupert, Edward R., and Robert D. Barns. *Invertebrate Zoology*. New York: Oxford University Press, 1994.

# Phylogenetics Systematics

This is a field of study that allows biologists to reconstruct a pattern of evolutionary events resulting in the distribution and diversity of present-day life. Achieving this goal requires classifying organisms into groups in a meaningful and universal manner. This classification is based on evolutionary events that occurred long before human civilization appeared on Earth. **Taxonomy** is the system used to name organisms based on their evolutionary relationships. A **taxon** is a hierarchical category used in the naming process, and taxa is the plural form of the word. The main taxa, in order of broadest to most specific designation, are: kingdom, phylum, class, order, family, genus, and species. For example, a domestic cat's kingdom is Animalia because it is an animal, its class is Mammalia because it is a mammal, and its genus and species are *Felis domesticus*. Phylogenetics refers to the study of

**fossil record** a collection of all known fossils

**mitochondrial DNA** DNA found within the mitochondria that control protein development in the mitochondria

**phyla** the broad, principal divisions of a kingdom

**salamander** a four-legged amphibian with an elongated body

**taxonomy** the science of classifying living organisms

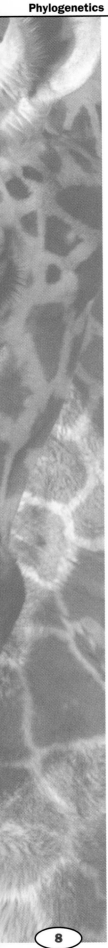

an organism's evolutionary history: when it first appeared on Earth, what it evolved from, where it lived, and when and why it went extinct (or survived). Systematics, then, refers to naming and organizing these biological taxa into meaningful relationships. For example, if two species of deer that are alive today are both thought to have evolved from a different species that subsequently went extinct, the taxonomic nomenclature (scientific name) of the deer should reflect that relationship.

Cladistics is an important tool for forming hypotheses about the relationships among organisms. Cladistics is a mechanism for providing a testable phylogenetic tree, a diagram representing the relationships of different organisms as a tree, with the oldest ancestors at the trunk of the tree and later descendants at the branch ends. The underlying assumption of cladistic analysis is that members of a single group are more closely related to each other than to members of a different group. When several organisms share a suite of features, they are grouped together because these shared features are likely to have belonged to a common ancestor of all the group members. When common features are thought to have this sort of evolutionary relevance, they are called "synapomorphies."

Conversely, those features that distinguish each member within a group from each other are called "apomorphies". These are derived characters, meaning that they evolved anew in the descendant and did not belong to the ancestor. As an example, both owls and sparrows have feathers and a beak because they share the synapomorphies of being birdlike; however, owls have very large eyes at the front of their head whereas sparrows have small eyes on either side of their head, and these are apomorphies. Cladistic analysis sums up the number of apomorphies and synapomorphies among different organisms and produces possible phylogenetic trees that minimize the apomorphies in particular groups. This is one method by which evolutionary relationships are estimated.

**morphological** related to the structure and form of an organism at any stage in its life history

The information that is used in cladistic analysis can be **morphological** or molecular. Morphological measurements are taken from fossils or from living animals. In fossil evidence, imprints of an organism or the fossilized organism itself provide evidence for the size and connectivity of hard body parts. Extant, or living, organisms make it possible also to measure the organism's soft parts, those that are unlikely to be fossilized. This is the most common type of cladistic study. Molecular evidence comes from comparing the genetic codes of extant species. Because DNA is thought to evolve at a constant rate, the **molecular clock** can be set at a particular, confidently estimated evolutionary event such as the divergence of **placental** from marsupial mammals. Then the amount of time since the divergence of two groups of organisms can be estimated based on the number of differences between their genetic codes. SEE ALSO PHYLOGENETIC RELATIONSHIPS OF MAJOR GROUPS.

**molecular clock** using the rate of mutation in DNA to determine when two genetic groups spilt off

**placental** having a structure through which a fetus obtains nutrients and oxygen from its mother while in the uterus

*Rebecca M. Steinberg*

**Bibliography**

Lincoln, Roger J., Geoffrey Allan Boxshall, and Paul F. Clark. *A Dictionary of Ecology, Evolution and Systematics.* Cambridge, U.K., and New York: Cambridge University Press, 1998.

Winston, Judith E. *Describing Species.* New York: Columbia University Press, 1999.

# Physiologist

Physiologists study the functions and activities of organisms—the way plants and animals are designed as well as how they interact with their environment. This includes functions and activities at the cellular and molecular level, both under normal and abnormal conditions. Physiologists may choose to specialize in any of the life processes, including growth, reproduction, aging, and metabolism, or the circulatory, nervous, or immune systems.

Notable physiologists include scientists such as American Dr. Matilda Brooks, who developed antidotes for cyanide and carbon monoxide poisoning. The Scottish physiologist Sir Charles Bell (1774–1842) described the central nervous system in human beings. A British physiologist, Edgar Adrian, shared a Nobel prize in 1932 for his work in determining the electrical nature of nerves and muscles, and later went on to codevelop the electroencephalograph which measures brain activity.

All physiologists require a background in physics and computer science with an emphasis on biological sciences such as microbiology, **ecology**, evolution, **genetics**, and **behavioral** biology. Physiologists work in either applied or basic research. Physiologists with a masters degree generally do applied research at companies interested in developing specific solutions to health problems or restoring the environment. They should be familiar with high-tech laboratory equipment such as electron microscopes, thermal cyclers, and nuclear magnetic resonance machines. They must also be able to communicate well with nonscientists.

Physiologists who pursue a Ph.D. spend additional time in laboratory research and in writing a dissertation. Frequently they go on for several years of post-doctorate work in their area of interest. Physiologists who specialize in basic research tend to work at universities, where they are funded by scientific grant money. Usually their work consists of doing original research, overseeing graduate students, and teaching. Unlike applied researchers, basic researchers are free to pursue knowledge for its own sake without the constraints of producing a practical product. This can lead to exciting discoveries, as they follow their curiosity into the mysteries of the world within and without.

*Nancy Weaver*

**ecology** the study of how organisms interact with their environment

**genetics** the branch of biology that studies heredity

**behavioral** relating to actions or a series of actions as a response to stimuli

**Bibliography**

*Occupational Outlook Handbook.* Washington, D.C.: U.S. Dept. of Labor/Bureau of Labor Statistics, 2000.

# Physiology

Physiology is the study of how living things function. It encompasses the most basic unit of living things, the cell, and the most complex organs and organ systems, such as the brain or endocrine system.

The word "physiology" was first used by the Greeks around 600 B.C.E. to describe a philosophical inquiry into the nature of things in general. Around the sixteenth century, the word began to be used with specific reference to the vital activities of healthy humans. By the nineteenth

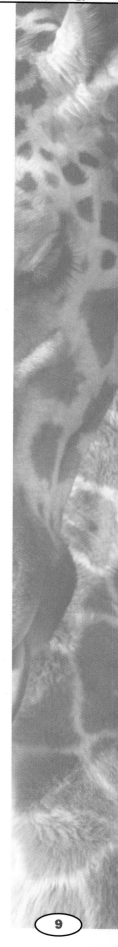

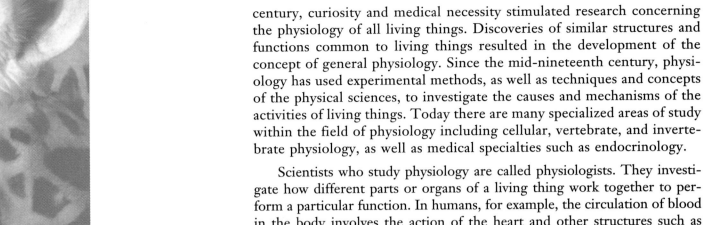

century, curiosity and medical necessity stimulated research concerning the physiology of all living things. Discoveries of similar structures and functions common to living things resulted in the development of the concept of general physiology. Since the mid-nineteenth century, physiology has used experimental methods, as well as techniques and concepts of the physical sciences, to investigate the causes and mechanisms of the activities of living things. Today there are many specialized areas of study within the field of physiology including cellular, vertebrate, and invertebrate physiology, as well as medical specialties such as endocrinology.

Scientists who study physiology are called physiologists. They investigate how different parts or organs of a living thing work together to perform a particular function. In humans, for example, the circulation of blood in the body involves the action of the heart and other structures such as veins, arteries, and capillaries. Special nerve centers known as nodes trigger the ventricles of the heart to contract in a predictable rhythm, which causes the blood to flow in and out of the heart. By learning how organs such as the heart function normally, physiologists (and physicians) can better understand what happens when organs function abnormally and learn how to treat them. In their studies, physiologists pay close attention to structure, information transfer, metabolism, regulation, and transport.

## Structure

The structures of living things are often related to their function. For example, the shape and structure of a bird's beak is related to how it uses the beak. Eagles have a large, sharp beak for ripping and tearing prey. Hummingbirds have long, slender beaks for sipping nectar from flowers. Physiologists often study and compare animal structures such as appendages (projecting structures or parts of an animal's body that are used in movement or for grasping objects) to determine similarities, differences, and evolutionary etiology (origin) among species.

## Information Transfer

Animals react quickly to external stimuli such as temperature change, touch, light, and vibration. Information from an organism's external environment is rapidly transferred to its internal environment. In **vertebrates**, nerve impulses initiated in sensory **neurons**, or nerve cells, are transferred to the center of the brain or **spinal cord**. Sensory neurons are nerve cells that transmit impulses from a receptor such as those in the eye or ear to a more central location in the nervous system. From the brain or spinal cord, impulses initiated in motor neurons (nerve cells that transmit impulses from a central area of the nervous system to an effector such as a muscle) are transferred to muscles and induce a reflex response. The brain and spinal cord receive incoming messages and initiate, or trigger, the motor neurons so that animals, including humans, can move.

## Metabolism

Metabolism is the processing of matter and energy within the cells, tissues, and organs of living organisms. There are four major questions to be answered in the study of metabolism: How do matter and energy move into

**vertebrates** animals with a backbone

**neurons** nerve cells

**spinal cord** a thick, whitish bundle of nerve tissue that extends from the base of the brain to the body

The reflex response is an automatic reaction, such as your knee jerking when the tendon below the knee cap is tapped; the impulse provoked by the tap, after travelling to the spinal cord, travels directly back to the leg muscle.

the cells? How are substances and forms of energy transformed within the cell? What function does each transformation serve? What controls and coordinates all the processes?

All animals require the atoms and molecules from food to build their bodies. Animals also require the energy released when chemical bonds are broken and new bonds are formed. This energy is required to do work and to maintain body temperature. Plants manufacture their own food by harvesting the energy of sunlight and storing the energy in the chemical bonds of carbohydrates, fats, and proteins. Animals cannot make their own food, so they obtain the energy of sunlight indirectly by eating plants or other animals.

The bodies of animals are composed of many different chemical compounds, including specialized proteins found in muscle tissue and in red blood cells. These proteins are not present in the food animals eat, so metabolism is the process of disassembling the proteins found in plant tissue into amino acids, then reassembling those amino acids in to the proteins that animals need.

Animals must use energy to assemble new molecules. Animals also require energy to pump blood, contract muscles, and maintain body temperature. This energy comes from the carbohydrates, **lipids**, and proteins animals eat. A complex series of reactions called the Kreb's cycle is the primary mechanism for the controlled release of energy from these molecules.

**lipids** fats and oils; organic compounds that are insoluble in water

## Regulation

Animals maintain their internal environments at a constant level. This process, called **homeostasis**, depends on the action of **hormones**. In humans, metabolic functions and hormone interactions expend energy and help to maintain a constant body temperature of 37°C (98.6°F). Comparative studies of neurosecretory cells, special nerve cells capable of secreting hormones, indicate that the cells are also important in the developmental and regulatory functions of most animals. In insects and **crustaceans**, hormones control the cycles of growth, **molting**, and development. By identifying the hormones that regulate these cycles in insects, scientists may be able to control insect pests by interfering with hormone production and thus, with the insect's processes of growth and development.

**homeostasis** a state of equilibrium in an animal's internal environment that maintains optimum conditions for life

**hormones** chemical signals secreted by glands that travel through the bloodstream to regulate the body's activities

**crustaceans** arthropods with hard shells and jointed bodies and appendages that mainly live in the water

**molting** the shedding of an exoskeleton as an animal grows so that a new, large exoskeleton can be secreted

## Transport

Most animals have a transport or circulatory system that involves the movement of oxygen and carbon dioxide through blood. In vertebrates and a few invertebrates, notably **annelids** and cephalopod **mollusks**, blood flows entirely in closed channels or vessels. In most other invertebrates, blood flows for part of its course in large sinuses (cavities or opening), or lacunae, and comes directly into contact with tissues. SEE ALSO BIOMECHANICS.

*Stephanie A. Lanoue*

**annelids** segmented worms

**mollusks** a large phylum of invertebrates that have soft, unsegmented bodies and usually have a hard shell and a muscular foot; examples are clams, oysters, mussels, and octopuses

**Bibliography**

Alexander, R. McNeil. *Animal Mechanics.* Seattle, WA: University of Washington Press, 1998.

# Plankton

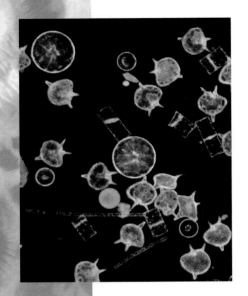

Marine plankton.

**eukaryotic cells** contain a membrane-bound nucleus and membrane-bound organelles

**flagella** cellular tails that allow the cell to move

**cilia** hair-like projections used for moving

**mitosis** a type of cell division that results in two identical daughter cells from a single parent cell

**meiosis** a specialized type of cell division that results in four sex cells or gametes that have half the genetic material of the parent cell

**ecosystems** self-sustaining collections of organisms and their environments

**abiotic factors** pertaining to nonliving environmental factors such as temperature, water, and nutrients

**autotrophs** organisms that make their own food

**heterotrophs** organisms that do not make their own food

Plankton (from the Greek word *planktos*, which means "wandering") are communities of mostly microscopic organisms that inhabit watery environments, from oceans to muddy regions. Some plankton drift passively or swim weakly near the surfaces of oceans, ponds, and lakes, while others exist as bottom-dwellers, attaching to rocks or creeping on the ground through sand and silt.

Plankton are classified under the kingdom Protista. During the genesis of protists, a true nucleus, as well as the other components of **eukaryotic cells** (mitochondria, chloroplasts, endoplasmic reticulum, Golgi bodies, 9+2 **flagella** and **cilia**, the functions **mitosis** and **meiosis**) arose. Thus these organisms are considered to be ancestral to plants, fungi and animals. While the majority of plankton are unicellular and therefore considered to be simple eukaryotic organisms, at the cellular level they are extremely complex. Plankton should be considered an organism in itself and not be compared to a single cell from a multicellular organism.

Despite their small size, plankton are the very basis for life in the earth's various **ecosystems**. An ecosystem is comprised of all the organisms living in a community and all **abiotic factors** with which the organisms interact.

The two main processes within an ecosystem are energy flow and chemical cycling. Energy enters most systems in the form of sunlight and is converted to chemical energy by **autotrophs**. The chemical energy is then passed to **heterotrophs** in organic compounds of food, and finally dissipates into the system as heat. **Trophic levels** are based on an organism's main source of nutrition. Autotrophs, also called primary producers, are generally photosynthetic organisms that use light energy to synthesize sugars and other organic compounds. Heterotrophs, or consumers, are supported by these photosynthetic organisms. The primary consumers are herbivores, who gain sustenance directly from autotrophs. Secondary consumers feed on the herbivores and tertiary consumers feed on the secondary ones. Those organisms that feed off of dead organisms are known as detritovores. An understanding of this pyramid within an ecosystem explains why the extent of photosynthetic activity determines the energy supply of the entire ecosystem.

Algae, as freshwater and marine phytoplankton and intertidal seaweeds, are responsible for nearly half of all photosynthetic production of organic material, rendering them extremely significant in the aquatic food webs where they support countless suspension-feeding and predatory animals. All algae, except prokaryotic cyanobacteria (formerly called blue-green algae), belong to the kingdom Protista. Algae all contain chlorophyll A, a primary pigment in cyanobacteria and plants, but differ in accessory pigments, which trap **wavelengths** of light to which chlorophyll A is not as sensitive. These accessory pigments include other chlorophylls (green), carotenoids (yellow-orange), xanthophylls (brown), and phycobilins (red and blue).

These differences in pigments point to different roles and effects of algae on the ecosystem. An overabundance of dinoflagellates (algae containing phycobilins) results in the blooming of red tides. When shellfish such as oysters feed on the dinoflagellates, they concentrate the algae along with toxic compounds released by the dinoflagellate cells. Because these toxins are dangerous to humans, collection of shellfish is restricted during red tides to reduce the risk of paralytic shellfish poisoning. Seaweed is the large marine algae that inhabits intertidal and subtidal zones of coastal waters. Coastal people, especially in Asia, harvest seaweed for food since it is high in iodine and other essential minerals. The brown alga laminaria is used in soups and the red alga porphyra is used to wrap sushi.

With so much dependent on the existence of plankton, the fight against water pollution aims to prevent not only the destruction of plankton, but of other species as well. SEE ALSO FOOD WEB; ZOOPLANKTON.

*Danielle Schnur*

### Bibliography

Campbell, Neil A. *Biology*, 3rd ed. Berkeley, CA: Benjamin/Cummings Publishing Company, Inc, 1993.

Odum, Howard T. *Systems Ecology: An Introduction.* New York: Wiley, 1984.

## Plate Tectonics *See Continental Drift.*

# Platyhelminthes

Animals in the phylum Platyhelminthes are called flatworms because they are flattened from head to tail. Flatworms share several features with more derived animal phyla. They are the most primitive group to exhibit **bilateral symmetry**. Flatworms have three embryonic tissue layers: ectoderm, **mesoderm**, and endoderm.

Animals within Platyhelminthes show more complexity than ancestral phyla, but are not as complex as more derived animal phyla. They are acoelomates, which means they do not have a body cavity. Platyhelminthes are unsegmented. They have muscles and a simple nervous system that includes a primitive brainlike structure which is formed from a thickening of the **ventral** nerve cords in the head region. They have a mouth, but no anus, and a primitive digestive cavity. They also have a primitive excretory system. They do not have a respiratory or circulatory system and are limited to simple **diffusion** for gas exchange. They can regenerate by **fission** as well as reproduce sexually, sometimes with complex life cycles passing through more than one host. Flatworms move about using cilia and by undulating movements of the whole body.

Almost all Platyhelminthes are aquatic, both fresh water and marine, but a few **terrestrial** species live in moist, warm areas. Species vary in size from microscopic to over 60 feet (20 meters) long for some tapeworms.

There are four major classes of Platyhelminthes and over twenty-five thousand species. Flatworms in the class Turbellaria are marine and freshwater free-living **scavengers**. The other three classes are parasitic and include some of the most harmful human parasites. The classes Trematoda, commonly called flukes, and Momogea are both **endoparasites** and **ectoparasites**. Momogea are parasites of aquatic vertebrates such as fish.

**trophic levels** divisions of species in an ecosystem by their main sources of nutrition

**wavelengths** distance between the peaks or crests of waves

**bilateral symmetry** characteristic of an animal that can be separated into two identical mirror image halves

**mesoderm** the middle layer of cells in embryonic cells

**ventral** the belly surface of an animal with bilateral symmetry

**diffusion** the movement of molecules from a region of higher concentration to a region of lower concentrations

**fission** dividing into two parts

**terrestrial** living on land

**scavengers** animals that feed on the remains of animals that they did not kill

**endoparasites** organisms that live inside other organisms and derive their nutrients directly from those organisms

**ectoparasites** organisms that live on the surfaces of other organisms and derive their nutrients directly from those organisms

Flatworms in the class Cestoda are endoparasites known as tapeworms. SEE ALSO PHYLOGENETIC RELATIONSHIPS OF MAJOR GROUPS

*Laura A. Higgins*

### Bibliography

Anderson, D. T., ed. *Invertebrate Zoology*. Oxford: Oxford University Press, 1998.

Barnes, Robert D. *Invertebrate Zoology*, 5th ed. New York: Saunders College Publishing, 1987.

Campbell, Neil A., Jane B. Reece, and Lawrence G. Mitchell. *Biology*, 5th ed. Menlo Park, CA: Addison Wesley Longman, Inc., 1999.

Purves, William K., Gordon H. Orians, H. Craig Heller, and David Sadava. *Life the Science of Biology*, 5th ed. Sunderland, MA: Sinauer Associates Inc. Publishers, 1998.

# Pleistocene

This most recent sequence of geologic time is somewhat complicated to describe in terms of animal science since there is much more information available from the fossil record. Scientifically, it is described as the period of time from 1.9 million year ago to 10,000 years ago. It is identified with a noticeable change in the animal fossils, which usually indicates some kind of extinction or massive change in the environment. It is difficult to summarize what happened from region to region since there was as much climatic variety then as there is today. However, the general consensus among scientists is that the beginning of the Pleistocene epoch began with an overall global cooling. This cooling was significant in that many cold-intolerant species disappeared and some new more resistant species appear in the fossil record.

Every geologic time period is defined by what scientists call a type section. A type section is a place that is considered to be the first discovered well-defined area in which evidence of a time-period shift, or difference between plant and animal communities, can be observed. In short, it is the first discovery of some important geological event characterized by a change in the kinds of species and populations of plant and animal fossils.

The type section for the Pleistocene was first proposed in 1839 by British geologist Charles Lyell after he examined a sequence (one of many layers) of rocks in southern Italy. He noticed that within and between the layers of rock, there was a distinct change between fossils of marine mollusks of warm-water species to fossils of species which were similar to modern cold-water species. After further investigation it was determined that this new set of geologic strata contained almost 70 percent living or historical species. Later studies in Europe by other geologists revealed that glaciation had occurred at about the same time as the strata in Italy were deposited. Eventually researchers pieced together evidence that indicated the Pleistocene was a time of great global cooling. During the epoch, immense glaciers and ice sheets occurred at the North and South Poles and at all high altitudes.

The Pleisocene was a relatively short span of geologic time that fluctuated between episodes of warming and cooling, but the general climate was very cold for much of the seas and regions of the continents.

The Pleistocene cooling had a tremendous effect on animal life on Earth; **faunas**, or ecological populations of animals, were severely disrupted or

Sir Charles Lyell (1797-1875) was a British geologist who opposed the catastrophic theory advanced at the time to account for great geologic changes. A proponent of uniformitarianism, Lyell is considered the father of modern geology.

**faunas** animals

| Era | Period | Epoch | Million Years Before Present |
|---|---|---|---|
| Cenozoic | Quartenary | Holocene | 0.01 |
| | | Pleistocene | 1.6 |
| | Tertiary | Pliocene | 5.3 |
| | | Miocene | 24 |
| | | Oligocene | 37 |
| | | Eocene | 58 |
| | | Paleocene | 66 |

The Pleistocene epoch and surrounding time periods.

eliminated altogether. Some species became extinct, while others flourished. Many new species have been identified as occurring around this time change and after. New species appeared both on land and at sea. As the ice sheets bound up more and more water, sea levels dropped. Land bridges appeared from beneath the sea, the most famous of which were the Bering land bridge and the land bridge between North and South America. Waves of animal migrations occurred on the continents.

## Animal Migrations to the Americas

Before the Pleistocene, North and South America contained their own distinctive sets of animals. Marsupials abounded in South America and the horse flourished in North America. With the emergence of the Panamanian land bridge between the two Americas, a great migration, or swapping of animal species, began. Marsupials (mammals with no **placenta** but with a pouch in which their young develop), sloths, and other animals such as glyptodonts, which looked like an armadillo, headed north. The proboscideans—including a group called the gomphotheres, with elongated lower jaws that looked like shovels—as well as mammoths, and mastodons, moved south. Relatives of modern horses, lions, camels, and wolves migrated after the gomphotheres.

The Bering land bridge that connects Russia and Alaska supported the invasion from Asia of animals such as the mammoth, deer and their relatives, and bison to the Americas. Perhaps the most influential animal to come across the land bridge was another mammal, *Homo sapiens*.

**placenta** a structure through which a fetus obtains nutrients and oxygen from its mother while in the uterus

## Animal Adaptations to Climate

Animals were also on the move in other parts of the world. The mammoths continued to migrate over Europe and Asia. The woolly mammoth developed a thick fur and began to graze in the spruce forests that bordered the ice. The rhinoceros also moved into Europe and central Asia and developed a coat of thick fur for surviving in the cold conditions. Its front horn grew to extreme lengths, reaching nearly a meter, and some researchers have suggested that legends describing the survivors of this species may have led to the myth of the unicorn. The massive and dangerous archaeocyonids, or bear dogs, were enormous predators whose bones are still found in caves today. In Europe, *Panthera leo spelaea*, a large species of cave lion, roamed the mountains in search of bison and other prey. In North America, *Smilodon*, the saber-toothed cats, traveled over the more warm and savanna-like regions of what is now the southwest United States.

A fossil beetle of the
Pleistocene period.

## Pleistocene Extinctions

**terrestrial** living on
land

**Terrestrial** invertebrates flourished and died with the fluctuating climate.
Records of species of snails show scientists how the climate cooled and
warmed throughout the period. Scientists identify these changes by the silt
deposits left by the advancing and retreating glaciers in the North. They
also use tree ring thicknesses (a part of dendrochronology) to determine
periods of dry or wet years. The best data for calculating oxygen and carbon dioxide isotopes (different numbers of electrons) come from ice cores
in Greenland. Small, single-celled marine animals called foraminifera were
able to secrete specialized shells. These small eukaryotes are extremely sensitive to temperature change and their tiny fossils leave an excellent record
of shifting climate for paleontologists to observe. The records of these animals, found in mud recovered from oil wells off coastal waters read like a
book of temperature fluctuations. When the water is warm a certain species
will abound. They die and sink to the mud where they are fossilized. When
the water is colder other species survive. These also leave their fossils in the
mud. An expert can read the sequence of fossils in the mud.

Curiously, amphibians and reptiles of the Pleistocene did not suffer the
extinctions that befell the mammals. Since these animals appear to be very
sensitive to climate today, it was assumed they would be affected by changes
in climate during the ice ages. Apparently they were not, and the fossils of
these animals did not change over time either in species or abundance. Their
geographic distribution may have changed as climates fluctuated, but there
are very few known extinctions of species. Birds also managed to survive.
Most of the birds that disappeared did so as a result of human interference
in recent times. The great moa of New Zealand was hunted to extinction
in the Holocene and is not considered a Pleistocene casualty.

The Pleistocene is famous for its extinctions rather than for its migrations. Some researchers believe the extinction event is not over and point

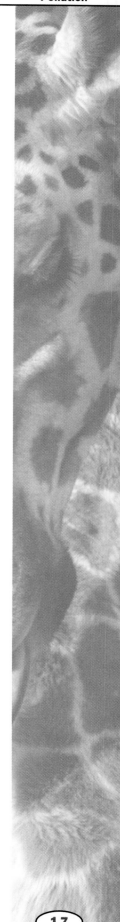

to the increasing list of endangered species throughout the world. Scientists are still not sure what caused the Pleistocene extinctions. Some hypothesize that many species could simply not tolerate the continuous climate fluctuations, others that temperatures were too cold. This is called the "Overchill Hypothesis." Other scientists note that wherever evidence of human migration is discovered, the large animals, or megafauna, disappear. These scientists believe that humans overhunted, and although not all the bison, deer, mammoths, and other large herbivores were killed for food, their disappearance led to the starvation of predators that relied on these animals. This is the "Overkill Hypothesis." Both hypotheses have merit, but still raise many questions. We may never know what caused the Pleistocene extinctions, but today, loss of habitat and increasing pollution are the most lethal killers of animal life on Earth. SEE ALSO PHYLOGENETIC RELATIONSHIPS OF MAJOR GROUPS.

*Brook Ellen Hall*

### Bibliography

Carroll, Robert. *Vertebrate Paleontology and Evolution.* New York: W. H. Freeman and Company, 1988.

Macdonald, D. *The Encyclopedia of Mammals.* New York: Facts on File Publications, 1987.

Martin, Paul, and Richard Klein. *Quaternary Extinctions: A Prehistoric Revolution.* Tucson, University of Arizona Press, 1989.

Woodburne, M. *Cenozoic Mammals of North America.* Berkeley: University of California Press, 1987.

# Pollution

Pollution can be defined as a change in the physical, chemical, or biological characteristics of the air, water, soil, or other parts of the environment that adversely affects the health, survival, or other activities of humans or other organisms. After pollution occurs, the polluted resource is no longer suitable for its intended use. Most pollutants are solid, liquid, or gaseous chemicals created as byproducts of the extractive or manufacturing industries. Pollution can also take the form of excessive heat, noise, light, or other electromagnetic radiation.

Pollution is also a complex political problem and affects people's lives in numerous ways other than health. Determining an acceptable level of pollution is often more of a political problem than a scientific problem. This is especially true when jobs are at stake, as when a manufacturing plant is forced to close or to modify its operations.

## Air Pollution

Air pollution is the contamination of air by unwanted gases, smoke particles, and other substances. Air pollution has been around since the beginning of the Industrial Revolution. In the United States, smoke pollution was at its worst in the first half of the twentieth century. The smoke was so thick in the winter in some industrial cities that street lights had to be left on all

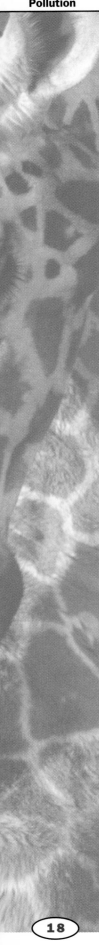

day. However, air pollution was considered a local problem. The burning of fossil fuels produced the most smoke, so some cities restricted the type of coal that could be burned to hard coal, which burns cleaner than soft coal. More efficient burners were installed and devices were attached to smokestacks to remove soot. Diesel locomotives replaced steam locomotives, which had burned coal or oil to heat the water to make the steam. These changes all led to a gradual reduction in smoke pollution during the last half of the twentieth century.

However, a new type of air pollution, smog, became a problem beginning in the 1940s. Smog is not the same as smoke pollution, although this was not immediately apparent. When Los Angeles had its first major smog attack, it became obvious that some phenomenon other than smoke was responsible. Los Angeles did not burn coal or oil to generate heat or electricity, yet its smog problem worsened. Scientists now know that smog is the result of the action of sunlight on unburned hydrocarbons and other compounds in the air. These unburned hydrocarbons come from the exhaust of motor vehicles with internal combustion engines.

Rapid industrial growth, urban and suburban development, dependence on motor vehicles, construction and operation of large facilities using fossil fuels for generating electricity, production of iron and steel, petrochemicals, and petroleum refining converted what had been a local problem into a regional or national problem. Many areas of the United States now suffer seasonal episodes of unhealthy air, including smog, haze, and **acid rain**. A wide range of pollutants currently poses an ecological threat in cities all over the United States and in other industrialized nations. Since pollution does not recognize national borders, the problem can spread around the world.

**acid rain** rain that is more acidic than non-polluted rain

## Costs of Air Pollution

The costs of air pollution in the United State are difficult to estimate. The economic benefits from the control of pollution are even harder to estimate. Business and industry bear the direct costs of compliance with regulations designed to control air pollution. Ultimately, consumers bear these costs through increased prices and reduced stock dividends. However, the economic benefits of cleaner air may be returned to the consumer in the form of lower health costs and longer-lasting and more reliable products.

## Types of Pollutants and Controls

There are six principal classes of air pollutants: particulate matter (soot), carbon monoxide, sulfur oxides, nitrogen oxides, unburned hydrocarbons, and ozone. Billions of metric tons of these compounds are discharged into the air each year.

**Particulates.** Suspended particulate matter such as soot and aerosols, are particulates. These particles of solids or liquids range in size from those that are visible as soot and smoke, to those that are so small they can only be seen through a microscope. Aerosols can remain suspended in the air for long periods and can be carried over great distances by the wind. Most particulates are produced by burning fossil fuels in power plants and other stationary sources. Controlling particulates usually involves washing, centrifugal separation, or electrostatic precipitation.

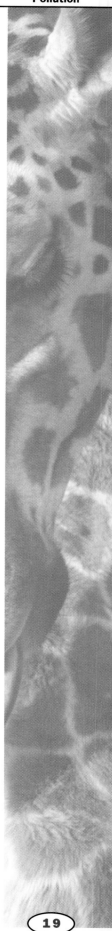

Particulates are harmful for a number of reasons. Some particles contribute to acid rain. Toxic materials such as lead or mercury can appear as particulates. However, the greatest health risk comes from breathing particulates. Some particles can lodge deep in the **lungs** and cause inflammation or chronic lung disease.

**lungs** sac-like, spongy organs where gas exchange takes place

**Carbon monoxide.** Carbon monoxide is a colorless, odorless, flammable, poisonous gas. The incomplete burning of carbon fuels produces carbon monoxide. Carbon monoxide comes largely from motor vehicles, with lesser amounts from other internal combustion engines such as those on lawnmowers and leafblowers, and from open fires and industrial processes. Carbon monoxide emissions can be controlled by more efficient burners or improved combustion chambers. Modern computer-controlled engines with catalytic converters successfully remove most carbon monoxide from automobile exhaust.

**Sulfur oxides.** Sulfur oxides are the major contributor to acid rain in the United States. Sulfur oxides include sulfur dioxide, sulfuric acid, and various sulfate compounds. Sulfur oxides are produced when fuel that contains sulfur is burned, or when metal ores are processed. Sulfur oxide emissions in the United States come primarily from plants that use fuels containing sulfur to generate electricity. The best way to reduce sulfur oxide emissions is to use fuels that naturally contain less than 1 percent sulfur, but these fuels are more expensive. Other techniques include removing sulfur from fuels and sulfur oxides from the combustion gases. Removing sulfur from stack gases after fuel is burned is difficult and expensive. However, the by-product, sulfuric acid, can be sold to recover some of the cost.

**Nitrogen oxides.** Nitrogen oxides are also contributors to acid rain and are a principal component of photochemical smog. Nitrogen oxides primarily result from the high-temperature combustion of gasoline or diesel in internal combustion engines. During combustion, nitrogen in the air chemically combines with oxygen to produce nitric oxide. Much of the nitric oxide is converted to nitrogen dioxide in a chemical reaction promoted by sunlight. Computer-controlled combustion and optimally designed combustion chambers can partially reduce the formation of nitrogen oxides. Special catalytic converters can combine nitric oxide with carbon monoxide and unburned hydrocarbons to produce nitrogen, carbon dioxide, and water.

**Unburned hydrocarbons.** Unburned hydrocarbons in air also represent wasted fuel. Gaseous hydrocarbons are not toxic at concentrations normally found in the atmosphere, but unburned hydrocarbons are a major contributor to the formation of ozone and smog. Catalytic converters on automobile engines have substantially reduced the emission of unburned hydrocarbons.

**Ozone.** Ozone is a form of oxygen in which the molecule contains three atoms instead of two. Ozone is beneficial when it is high in the atmosphere, but near the surface, ozone can damage rubber and paint as well as damage lung tissue. Ozone is a constituent of smog, and is produced from the reaction of nitrogen oxides with gaseous hydrocarbons in the presence of sunlight. A small amount of ozone is also produced by lightning storms. The control of ozone and other photochemical oxidants depends on the effective control of both nitrogen oxides and gaseous hydrocarbons.

## Water Pollution

Water pollution is caused by any chemical, physical, or biological substance that affects the natural condition of water or its intended use. We rarely stop to think about how important a reliable and safe water supply is until it is restricted or damaged. Water pollutants are produced primarily by the activities of humans. Our fresh water supply is under worldwide threat from pollution.

Water in lakes and rivers (surface water) throughout the world must satisfy a wide variety of different needs. Some of these needs partially conflict with others:

- The public wants lakes and rivers preserved in their natural state.
- Lakes and rivers must support a healthy population of fish and wildlife.
- Surface water must be safe for recreational uses such as swimming.
- Many localities depend on surface water for a safe drinking water supply.
- Surface water must be safe for agricultural use.
- Surface water must accommodate a variety of industrial purposes
- Surface water is used to generate power or cool power plants.
- Surface water is counted upon to dilute and transport human and industrial waste.

Because of the complex factors involved, there is no precise definition of water pollution. Instead, the intended use of the water must be considered. Once the intended use of the water is specified, pollutants can be grouped as not permissible, as undesirable and objectionable, as permissible but not necessarily desirable, or as desirable. For example, if water is to be used for wildlife support and enhancement, toxic compounds are not permissible, but oxygen is desirable. If the water is to be converted to steam in a power plant, some toxic materials might be desirable (because they reduce zebra mussel infestations), while excess oxygen that could corrode equipment would be undesirable.

Another method of classifying water pollutants is to distinguish between pollutants that are not altered by the biological processes occurring in natural waters, and those that will eventually break down into other, perhaps less objectionable, compounds. Inorganic chemicals are diluted by water but do not chemically change. Industrial waste often contains this sort of pollutant (for example, mercury). On the other hand, domestic sewage can be converted into inorganic materials, such as bicarbonates, sulfates, and phosphates, by the action of bacteria and other microorganisms in the water. If the water is not too heavily laden with waste, bacteria can break down the waste to safe levels.

Until early in the twentieth century, efforts to control water pollution were directed toward eliminating potential disease-causing organisms, such as typhoid. This led to treatment plants to provide safe drinking water and measures to enhance the natural biological activity of streams and rivers in order to assimilate and break down waste. By the middle of the twentieth

In the United States, concern for the natural condition of water was first expressed in the 1899 Rivers and Harbors Appropriation Act. The measure made it illegal to dump waste into waters used by any kind of vessels, except by special permission.

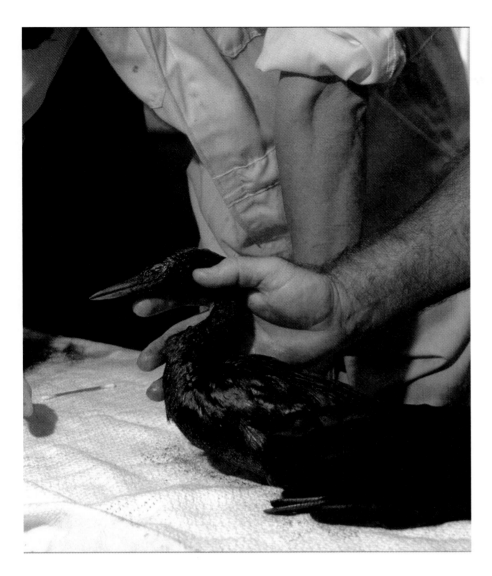

Two men attempt to clean a bird caught up in the oil spilled into the Persian Gulf during the 1991 Gulf War. Feathers soaked with oil lose their waterproofing characteristics—the feature birds need to stay afloat in the water.

century, the focus had shifted to the treatment of chemical pollutants not removed by conventional water-treatment methods.

By the middle of the twentieth century, the situation was changing. Rapidly growing urban areas generated large quantities of waste that had to be processed. Increased manufacturing capacity greatly increased the amount and variety of industrial waste. Commercial fertilizers and pesticides created many new pollution problems. Sewer systems were often unable to keep pace with rapid urban growth. Today, virtually every body of water on Earth has some degree of water pollution. Even the oceans, which were once thought to be able to absorb an unlimited amount of waste, are now showing significant stress due to pollution.

## Major Water Pollutants

Water is considered polluted if it contains an amount of any substance that renders the water unsuitable for a particular purpose. The list of substances that may pollute water is very long, and only a few major pollutants can be discussed here.

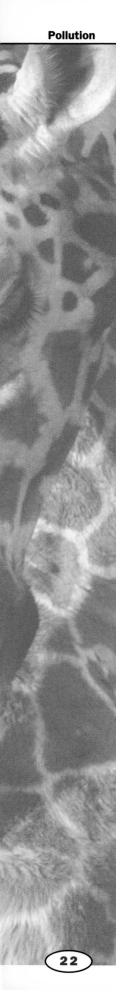

**Organic waste.** Organic waste comes from domestic sewage, agricultural runoff, feedlot operations, and industrial waste of animal and plant origin, such as from a paper mill. Domestic sewage is the largest and most widespread source of organic waste. Industrial organic waste tends to occur in larger quantities at fewer locations. Industries that make food and paper (and wood pulp) produce the largest amounts of industrial organic waste.

Bacteria can efficiently break down organic waste. However, bacterial action also removes oxygen from the water. Because fishes and other forms of **aquatic** life depend on dissolved oxygen, the bacterial action necessary to break down the waste damages the aquatic environment. If organic waste consumes oxygen at a rate greater than it can be replenished, then anaerobic bacteria dominate the decay process. Anaerobic decomposition by bacteria is smelly and aesthetically unpleasant.

**aquatic** living in water

**Plant nutrients.** Nitrogen and phosphorus are the two main plant nutrients acting as polluting agents. If plant nutrients get into water, they stimulate the growth of algae and other water plants. When these plants die and decay, they consume oxygen, just like any other organic waste. The excess plant growth caused by fertilization and subsequent build up of dead plant matter is called eutrophication. If the oxygen level drops even a small amount, desirable species of fishes, such as trout and bass, will be replaced by less desirable species such as carp and catfish. If the oxygen level drops low enough, all species of fishes, crayfishes, shrimp, and other organisms may die.

**Synthetic organic chemicals.** The water pollution problem causing the greatest concern is the ever-increasing variety of new chemical compounds. Often new compounds are developed and old ones abandoned before their environmental impact is known. Some of these compounds will remain in water for decades, or longer. The presence of these synthetic chemicals adversely affect fishes and other aquatic life. Many researchers think that some synthetic chemicals mimic natural hormones, disrupting growth and reproductive cycles in affected populations.

**Inorganic chemicals.** Inorganic chemicals such as mercury, nitrates, phosphates, and other compounds may also enter surface water. Many of these chemicals destroy fish and aquatic life, cause excessive hardness of water supply, and corrode machinery. This adds to the cost of water treatment.

Mercury pollution has been recognized as a serious, chronic, and widespread danger in many waterways. Even very small amounts of mercury can cause serious physiological effects or even death. Because mercury is not normally found in food or water, no organisms have developed the ability to process and excrete mercury. So it collects in tissues until toxic levels are reached. Mercury also undergoes **biomagnification**. Organisms at higher trophic levels consume a large number of organisms at lower trophic levels. The concentration of mercury becomes progressively higher at higher trophic levels. Predators at the highest trophic levels can accumulate dangerously high levels of mercury in this way.

**biomagnification**
increasing levels of toxic chemicals through each trophic level of a food chain

**Radioactive materials.** Radioactive materials are a recent addition to the list of potential water pollutants. Radioactive waste comes from the mining and processing of radioactive ores, from the refining of radioactive materials, from the industrial, medical, and research uses of radioactive materials, and from nuclear-powered reactors. Some radioactive waste still remains from the atmospheric testing of nuclear weapons in the 1940s and 1950s. The two most common radioactive materials found in water are strontium-90 and radium-226.

**Oil.**    Oil pollution can enter water through bilge flushing, from accidental or deliberate discharge from ships, or from accidental spills of crude oil during transport. Some experts estimate that 1.5 million tons of oil are spilled into the oceans each year. Water polluted by oil greatly damages aquatic life and other wildlife such as birds that depend on the water for food and nesting areas. Waterfowl alighting on oil-covered waters usually become so oil soaked that they are unable to fly. One speck of oil on a bird's feathers can poison the bird if it ingests the oil while preening. Oil destroys much of the aquatic life of oceans. It is particularly damaging to shellfish and other **filter feeders**. It also damages the small shrimp and other organisms that serve as food for larger fish.

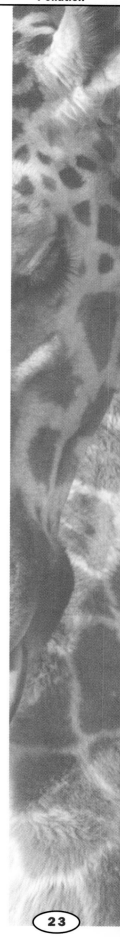

**filter feeders** animals that strain small food particles out of water

**Thermal pollution.**    Thermal pollution is caused by the release of heat into the water or air. Electric power plants are a major source of thermal pollution, since they convert only about one-third of fuel energy into electricity. The remaining heat is discharged to the local environment as heated water or air. This can alter the ecological balance of a large area. For example, if warm water is discharged into a lake, the warm water will not be able to dissolve as much oxygen. This may result in more desirable species of fishes being replaced by less desirable species.

## Land and Soil Pollution

One of the miracles of technology in the late twentieth century has been the extraordinary ability of agriculture to increase the productivity of croplands to previously unheard of levels. However, this increased productivity requires the heavy use of pesticides and fertilizer. Most pesticides today are designed to decompose very quickly into harmless compounds. Thousands of pesticides are currently in use, and in most cases their agricultural value balances their risks. However, many scientists think that we may be in an unbreakable cycle of having to continually develop new and more potent pesticides to overcome pests that are resistant to older pesticides.

## Noise Pollution

Noise pollution is a recently identified source of environmental degradation. The hearing apparatus of living things is sensitive to certain frequency ranges and sound intensities. Sound intensities are measured in decibels. A sound at or above the 120 decibel level is painful and can injure the ear. Noise pollution is present even in the open ocean. Researchers have shown that whales communicate over great distances using low frequency sound waves. Unfortunately, the noise generated by engines and screws of ships falls in the same frequency range and can interfere with the whale's communication.

## Light Pollution

Professional and amateur astronomers have recently identified a problem that did not exist a generation ago: light pollution. This form of pollution

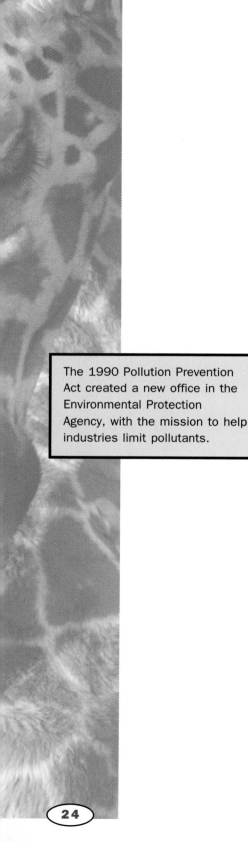

has spread so widely in the last few decades that many people are only able to see a few of the brightest stars. From many of our big cities, no stars at all are visible. Light pollution is not an inevitable consequence of making our streets and neighborhoods safer. Most light pollution comes from wasted light. At least 75 percent of the sky glow in most cities comes from poorly designed or improperly installed light fixtures. According to a study conducted by the International Dark-Sky Association in 1997, about $1.5 billion per year is spent on wasted light that does nothing to improve security or safety.

## Efforts to Control Pollution

Because pollution does not recognize national borders, the solution to many pollution problems requires cooperation at regional, national, and international levels. For example, smokestacks of coal-burning power plants in the United States cause some of the acid rain that falls in Canada.

In the United States, the Environmental Protection Agency (EPA) is charged with enforcing the many and complex laws, rules, executive orders, and agency regulations regarding the environment. The EPA came into being in 1969 with the passage of the National Environmental Policy Act (NEPA). This act also required the filing of environmental impact statements. Almost all government agencies and many businesses are required by law to file these statements, which state the potential harmful environmental effects of such activities as opening new factories, building dams, and drilling new oil wells.

Additional laws to protect the environment were passed in the 1970s and 1980s, including the Clean Air Act, the Safe Drinking Water Act, and the Comprehensive Environmental Response, Compensation, and Liability Act, known as Superfund. Then-president George Bush signed the Clean Air Act of 1990. This new law called for substantial reductions in emissions of all types. The act also added to the list of potentially toxic chemicals that must be monitored by the EPA. The pace of new environmental legislation has waned since 1990, but congress enacted the Food Quality Protection Act in 1996 and the Chemical Safety Information, Site Security and Fuels Regulatory Relief Act in 1999. SEE ALSO HABITAT LOSS; TROPHIC LEVEL.

*Elliot Richmond*

The 1990 Pollution Prevention Act created a new office in the Environmental Protection Agency, with the mission to help industries limit pollutants.

**Bibliography**

Art, Henry W., ed. *The Dictionary of Ecology and Environmental Science*. New York: Holt, 1993.

Bowler, Peter J. *Norton History of Environmental Sciences*. New York: Norton, 1993.

Caplan, Ruth. *Our Earth, Ourselves: The Action-Oriented Guide to Help You Protect and Preserve Our Planet*. Toronto, Canada: Bantam, 1990.

Carson, Rachel. *Silent Spring*. Boston: Houghton Mifflin, 1994.

Chiras, Daniel D. *Environmental Science: Action for a Sustainable Future*, 4th ed. Menlo Park, CA: Benjamin-Cummings Publishing Company, 1993.

Franck, Irene M., and David Brownstone. *The Green Encyclopedia*. New York: Prentice Hall General Reference, 1992.

Harms, Valerie. *The National Audubon Society Almanac of the Environment*. New York: Putnam Publishing, 1994.

Luoma, Jon R. *The Air Around Us: An Air Pollution Primer*. Raleigh, NC: Acid Rain Foundation, 1989.

Miller Jr., G. Tyler. *Living in the Environment,* 6th ed. Belmont, CA: Wadsworth Publishing Company, 1990.

Mott, Lawrie, and Karen Snyder. *Pesticide Alert: A Guide to Pesticides in Fruits and Vegetables.* San Francisco: Sierra Club Books, 1997.

Newton, David. *Taking a Stand Against Environmental Pollution.* Danbury, CA: Franklin Watts Incorporated, 1990.

Rubin, Charles T. *The Green Crusade: Rethinking the Roots of Environmentalism.* New York: Free Press, 1994.

# Population Dynamics

A population describes a group of individuals of the same species occupying a specific area at a specific time. Some characteristics of populations that are of interest to biologists include the **population density**, the **birthrate**, and the **death rate**. If there is immigration into the population, or emigration out of it, then the immigration rate and emigration rate are also of interest. Together, these **population parameters,** or characteristics, describe how the population density changes over time. The ways in which population densities fluctuate—increasing, decreasing, or both over time—is the subject of population dynamics.

Population density measures the number of individuals per unit area, for example, the number of deer per square kilometer. Although this is straightforward in theory, determining population densities for many species can be challenging in practice.

## Measuring Population Density

One way to measure population density is simply to count all the individuals. This, however, can be laborious. Alternatively, good estimates of population density can often be obtained via the quadrat method. In the quadrat method, all the individuals of a given species are counted in some subplot of the total area. Then that data is used to figure out what the total number of individuals across the entire habitat should be.

The quadrat method is particularly suited to measuring the population densities of species that are fairly uniformly distributed over the habitat. For example, it has been used to determine the population density of soil species such as nematode worms. It is also commonly used to measure the population density of plants.

For more mobile organisms, the **capture-recapture method** may be used. With this technique, a number of individuals are captured, marked, and released. After some time has passed, enough time to allow for the mixing of the population, a second set of individuals is captured. The total population size may be estimated by looking at the proportion of individuals in the second capture set that are marked. Obviously, this method works only if one can expect individuals in the population to move around a lot and to mix. It would not work, for example, in territorial species, where individuals tend to remain near their territories.

The birthrate of a population describes the number of new individuals produced in that population per unit time. The death rate, also called mortality rate, describes the number of individuals who die in a population per

**population density** the number of individuals of one species that live in an given area

**birthrate** a ratio of the number of births in an area in a year to the total population of the area

**death rate** a ratio of the number of deaths in an area in a year to the total population of the area

**population parameters** a quantity that is constant for a particular distribution of a population but varies for the other distrubutions

**capture-recapture method** a method of estimating populations by capturing a number of individuals, marking and releasing them, and then seeing what percentage of newly captured individuals are captured again

King penguins survey the scene at St. Andrews Bay, South Georgia Island.

unit time. The immigration rate is the number of individuals who move into a population from a different area per unit time. The emigration rates describe the numbers of individuals who migrate out of the population per unit time.

The values of these four population parameters allow us to determine whether a population will increase or decrease in size. The "intrinsic rate of increase $r$" of a population is defined as $r = (birth\ rate\ +\ immigration\ rate) - (death\ rate\ +\ emigration\ rate)$.

If $r$ is positive, then more individuals will be added to the population than lost from it. Consequently, the population will increase in size. If $r$ is negative, more individuals will be lost from the population than are being added to it, so the population will decrease in size. If $r$ is exactly zero, then the population size is stable and does not change. A population whose density is not changing is said to be at **equilibrium**.

**equilibrium** a state of balance

## Population Models

We will now examine a series of population models, each of which is applicable to different environmental circumstances. We will also consider how closely population data from laboratory experiments and from studies of natural populations in the wild fit these models.

**Exponential growth.**    The first and most basic model of population dynamics assumes that an environment has unlimited resources and can support an unlimited number of individuals. Although this assumption is clearly unrealistic in many circumstances, there are situations in which resources are in fact plentiful enough so that this model is applicable. Under these circumstances, the rate of growth of the population is constant and equal to the intrinsic rate of increase $r$. This is also known as **exponential growth**.

**exponential growth** a population growing at the fastest possible rate under ideal conditions

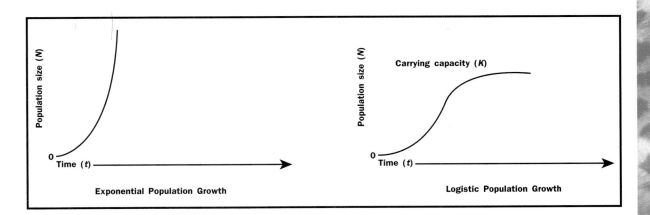

Population growth predicted by exponential and logistical models.

What happens to population size over time under exponential growth? If $r$ is negative, the population declines quickly to extinction. However, if $r$ is positive, the population increases in size, slowly at first and then ever more quickly. Exponential growth is also known as "J-shaped growth" because the shape of the curve of population size over time resembles the letter "J." Also, because the rate of growth of the population is constant, and does not depend on population density, exponential growth is also called "density-independent growth." Exponential growth is often seen in small populations, which are likely to experience abundant resources. J-shaped growth is not sustainable however, and a population crash is ultimately inevitable.

There are numerous species that do in fact go through cycles of exponential growth followed by population crashes. A classic example of exponential growth resulted from the introduction of reindeer on the small island of Saint Paul, off the coast of Alaska. This reindeer population increased from an initial twenty-five individuals to a staggering two thousand individuals in twenty-seven years. However, after exhausting their food supply of lichens, the population crashed to only eight. A similar pattern was seen following the introduction of reindeer on Saint Matthew Island, also off the Alaskan coast, some years later. Over the course of history, human population growth has also been J-shaped.

**Logistic growth.** A different model of population increase is called **logistic growth**. Logistic growth is also called "S-shaped growth" because the curve describing population density over time is S-shaped. In S-shaped growth, the rate of growth of a population depends on the population's density. When the population size is small, the rate of growth is high. As population density increases, however, the rate of growth slows. Finally, when the population density reaches a certain point, the population stops growing and starts to decrease in size. Because the rate of growth of the population depends on the density of the population, logistic growth is also described as "density-dependent growth".

Under logistic growth, an examination of population size over time shows that, like J-shaped growth, population size increases slowly at first, then more quickly. Unlike exponential growth, however, this increase does not continue. Instead, growth slows and the population comes to a stable equilibrium at a fixed, maximum population density. This fixed maximum is called the **carrying capacity**, and represents the maximum number of in-

**logistic growth** in a population showing exponential growth the individuals are not limited by food or disease

**carrying capacity** the maximum population that can be supported by the resources

dividuals that can be supported by the resources available in the given habitat. Carrying capacity is denoted by the variable $K$.

The fact that the carrying capacity represents a stable equilibrium for a population means that if individuals are added to a population above and beyond the carrying capacity, population size will decrease until it returns to $K$. On the other hand, if a population is smaller than the carrying capacity, it will increase in size until it reaches that carrying capacity. Note, however, that the carrying capacity may change over time. $K$ depends on a wealth of factors, including both **abiotic** conditions and the impact of other biological organisms.

**abiotic** nonliving parts of the environment

Logistic growth provides an accurate picture of the population dynamics of many species. It has been produced in laboratory situations in single-celled organisms and in fruit flies, often when populations are maintained in a limited space under constant environmental conditions.

Perhaps surprisingly, however, there are fewer examples of logistic growth in natural populations. This may be because the model assumes that the reaction of population growth to population density (that is, that population growth slows with greater and greater population densities, and that populations actually decrease in size when density is above the carrying capacity) is instantaneous. In actuality, there is almost always a time lag before the effects of high population density are felt. The time lag may also explain why it is easier to obtain logistic growth patterns in the laboratory, since most of the species used in laboratory experiments have fairly simple life cycles in which reproduction is comparatively rapid.

Biological species are sometimes placed on a continuum between $r$-selected and $k$-selected, depending on whether their population dynamics tend to correspond more to exponential or logistic growth. In $r$-selected species, there tend to be dramatic fluctuations, including periods of exponential growth followed by population crashes. These species are particularly suited to taking advantage of brief periods of great resource abundance, and are specialized for rapid growth and reproduction along with good capabilities for dispersing.

In $k$-selected species, population density is more stable, often because these species occupy fairly stable habitats. Because $k$-selected species exist at densities close to the carrying capacity of the environment, there is tremendous competition between individuals of the same species for limited resources. Consequently, $k$-selected individuals often have traits that maximize their competitive ability. Numerous biological traits are correlated to these two **life history strategies**.

**life history strategies** methods used to overcome pressures for foraging and breeding

**Lotka-Volterra models.**  Up to now we have been focusing on the population dynamics of a single species in isolation. The roles of competing species, potential prey items, and potential predators are included in the logistic model of growth only in that they affect the carrying capacity of the environment. However, it is also possible directly to consider between-species interactions in population dynamics models. Two that have been studied extensively are the Lotka-Volterra models, one for competition between two species and the other for interactions between predators and prey.

Competition describes a situation in which populations of two species utilize a resource that is in short supply. The Lotka-Volterra models of the population dynamics of competition show that there are two possible results: either the two competing species are able to **coexist**, or one species drives the other to extinction. These models have been tested thoroughly in the laboratory, often with competing yeasts or grain beetles.

Many examples of competitive elimination were observed in lab experiments. A species that survived fine in isolation would decline and then go extinct when another species was introduced into the same environment. Coexistence between two species was also produced in the laboratory. Interestingly, these experiments showed that the outcome of competition experiments depended greatly on the precise environmental circumstances provided. Slight changes in the environment—for example, in temperature—often affected the outcome in competitions between yeasts.

Studies in natural populations have shown that competition is fairly common. For example, the removal of one species often causes the abundance of species that share the same resources to increase. Another important result that has been derived from the Lotka-Volterra competition equations is that two species can never share the same **niche**. If they use resources in exactly the same way, one will inevitably drive the other to extinction. This is called the **competitive exclusion principle**. The Lotka-Volterra models for the dynamics of interacting predator and prey populations yields four possible results. First, predator and prey populations may both reach stable equilibrium points. Second, predators and prey may each have never-ending, oscillating (alternating) cycles of increase and decrease. Third, the predator species can go extinct, leaving the prey species to achieve a stable population density equal to its carrying capacity. Fourth, the predator can drive the prey to extinction and then go extinct itself because of starvation.

As with competition dynamics, biologists have tried to produce each of these effects in laboratory settings. One interesting result revealed in these experiments was that with fairly simple, limited environments, the predator would always eliminate the prey, and then starve to death. The persistence of both predator and prey species seemed to be dependent on living in a fairly complex environment, including hiding places for the prey.

In natural populations, studies of predator-prey interactions have involved predator removal experiments. Perhaps surprisingly, it has often proven difficult to demonstrate conclusively that predators limit prey density. This may be because in many predator-prey systems, the predators focus on old, sick, or weak individuals. However, one convincing example of a predator limiting prey density involved the removal of dingoes in parts of Australia. In these areas, the density of kangaroos skyrocketed after the removal of the predators.

Continuing oscillations between predators and prey do not appear to be common in natural populations. However, there is one example of oscillations in the populations of the Canada lynx and its prey species, the snowshoe hare. There are peaks of abundance of both species approximately every ten years. SEE ALSO POPULATIONS.

*Jennifer Yeh*

**coexist** to both be present

**niche** how an organism uses the biotic and abiotic resources of its environment

**competitive exclusion principle** the concept that when populations of two different species compete for the same limited resources, one species will use the resources more efficiently and have a reproductive edge and eventually eliminate the other species

**Bibliography**

Curtis, Helena. *Biology.* New York: Worth Publishers, 1989.

Gould, James L., and William T. Keeton, and Carol Grant Gould. *Biological Science,* 6th ed. New York: W. W. Norton & Co., 1996.

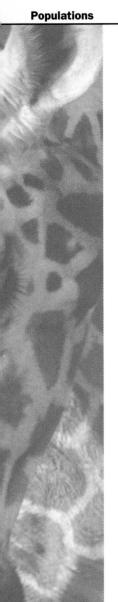

Krebs, Charles J. *Ecology: The Experimental Analysis of Distribution and Abundance.* New York: Harper Collins College Publishers, 1994.

Murray, Bertram G., Jr. *Population Dynamics: Alternative Models.* New York: Academic Press, 1979.

Pianka, Eric R. *Evolutionary Ecology.* New York: Addison Wesley Longman, 2000.

Ricklefs, Robert E., and Gary L. Miller. *Ecology,* 4th ed. New York: W. H. Freeman, 2000.

Soloman, Maurice E. *Population Dynamics.* London: Edward Arnold, 1969.

# Populations

The study of populations and population ecology is a growing field of biology. Plants and animals are studied both singly and in relationship to one another. Factors that affect population growth and overall health are constantly being sought and analyzed.

Animal populations are a bit easier to discuss since the genetic basis from which animal populations arise is not as complicated as the genetic basis of plants. Animal populations are more constrained by genetic variation than are plants. No haploid or **polyploid** animal exists or reproduces. Population dynamics (growth, death, and reproduction rate) are more easily explained in animals and many models, or predictions of population success, can be used to examine and learn from animal populations. Insects, in particular, provide a wealth of interesting population models since they tend to reproduce rapidly and in high numbers and their life cycles are fairly short.

**polyploid** having three or more sets of chromosomes

A great deal of healthy debate exists regarding the definition of a population. Not all species fit neatly into any one definition. In general, a population is described as a group of individuals of a species that lives in a particular geographic area. It is a sexually reproducing species in which individuals add to the continued growth or sustenance of the population. The sexually reproducing component of the definition is critical in that many endangered populations are at risk simply because they are not reproducing effectively enough to sustain their populations.

An example of this was found in the remnant population of the California condor. Its numbers in the wild had dwindled to approximately twenty individuals. The condors were not successfully reproducing in their natural habitat. Each year the eggs were infertile or crushed and no offspring were being reared. The population continued to dwindle as the older birds died and no young birds were born to replace them. Eventually, they were all captured and artificially bred in hopes of preserving the species. The population has increased in captivity, but California condors still face the challenge of becoming a viable biological population in the wild.

Raising captive baby condors requires all sorts of skills. After feeding the young birds had become an issue, their caretakers started to wear hand puppets designed in the image of the head of a female California condor. The babies quickly adjusted to taking their food from the mouth of their fabricated "mother." This gimmick made a difference in the number of babies surviving in captivity.

A population can exist over a broad geographic expanse, such as the North American continent or even the Earth, but it can also exist in one small pond. Many different populations of desert pupfish live only in their own small pond throughout the Mojave Desert. Populations of mites or parasites may live on one specific host or they may prefer one area of the host organism. For instance, a population of tapeworms may live only in the intestinal tract of the host while a louse population exists in the external environment. Most often a population for biological study is one in which a distinct geographical range can be assessed, such as a valley, plain, or forest, and a definite **genotype** can be identified.

A great deal of genetic variation exists within a population and the predictions of how this variation may be expressed is the fascinating work of population geneticists. Mathematics is used to help predict how a trait will move through a population or how a population will respond to an environmental pressure. Mathematical models that help scientists study population response to internal (genetic) and external (environmental) pressures are predictive only and are never entirely correct.

**genotype** the genetic makeup of an organism

It is the very nature of scientific models to be incorrect. No human or computer can ever account for all the existing variables or potential variables that may affect a population. How would one anticipate an intensely cold and violent storm three years from now? How would it be possible to predict a specific genetic mutation? None of this can be done, but all natural variables eventually affect populations in some way or another.

## Environmental Variables

All populations are ultimately controlled by the carrying capacity of their environment. The carrying capacity is the sum of all resources needed by a specific species in order to survive. The abundance of food is a major control. When food is short, young, old, and unfit members will die.

If two or more species are competing for one food source, additional pressure is placed on both populations. If two species of birds rely heavily on a certain insect for protein and nutrition, the availability and abundance of that insect is crucial. Any reduction in the population of insects will result in a loss of species from either bird population. There may not be enough food to feed and rear the young. Older birds unable to fly long distances for alternate food will also perish.

Food is not the only limiting factor that affects carrying capacity. Shelter and places to safely rear young are also part of the limits for many populations. Ground squirrels often rear their young underground or in rock shelters. This protects the young from predators, such as foxes or birds of prey. If there is a lack of adequate shelter the young are in peril. The rock squirrel lives in the crevices of rocks. If there are no rocks in a particular feeding area the squirrel may be able to eat out in the open or under a tree, but hiding from snakes and birds would be difficult under such circumstances. Rock squirrels in an area without the necessary physical habitat would be unable to successfully raise young to replenish the next generation.

Climate variables can also affect carrying capacity. If normal weather conditions change significantly over a period of time, certain plant species may not survive. Animals living on those plants will also perish. Pollution may also prevent populations from surviving; water and food may become toxic and make the environment unsuitable for existing populations.

Ultimately, the carrying capacity of an area affects the growth of a population since the numbers of species will not grow if there are not enough resources for the individuals to survive. The population will be regulated by the availability of food and other resources. Numbers of individuals will not increase, but in all likelihood remain the same.

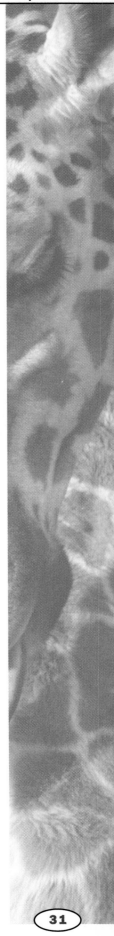

The health of this group of minnows, also known as shiners, can be greatly dependant upon environmental conditions.

## Population Regulation

Population regulation is achieved through several mechanisms. The environment is always at the top of the list of population regulation factors for reasons just mentioned. Another factor is the number of predators in any given area. If an animal that feeds on other animals is maintaining a stable population, it will only eat so many prey species every year. If, however, the prey species begins to flourish and increase in numbers, there will be more food for the predator. The predator population will also become more successful and increase in numbers. Eventually the predators will eat all the available prey and they will have reached the limits or carrying capacity of their environment.

The classic example of this type of population regulation is the Canadian lynx and the snowshoe hare. With regular cyclicity, hare populations increase in their habitats. As they rear more and more young, the available food for the Canadian lynx increases. With readily available food, the lynx is also more successful at raising young, so its population grows in response. Eventually the limit of how many snowshoe hares can exist in a particular region is reached. The lynx eat more and more hares, thus regulating and even reducing the population of hares. As the snowshoe hare population declines so does the population of Canadian lynx. Each population is regulating the growth of the other.

## Density Dependence and Independence

Density-dependent and density-independent populations are the focus of many scientific models. A density-dependent population is one in which the number of individuals in the population is dependent on a variety of factors including genetic variability and the carrying capacity of its environment. In a density-dependent population the ability to find a mate is critical. When population numbers are low this may be a critical factor in the survival of the species. If no mates are found during a season there will be no offspring.

Another effect of density-dependency is **intraspecific** (within the species) competition. When members of a population are all competing for food resources or adequate habitat, the less fit (unhealthy, young, or old) members will lose out and perish. As a population grows, this type of intraspecific competition serves as a self-regulating mechanism, eliminating many members of the population.

**intraspecific** involving members of the same species

Density-independent populations are those that are regulated by catastrophic or unusual events. Hurricanes constantly provide population controls on the coasts of North America or neighboring islands. Winds bring down trees where birds and other animals find shelter. Water drowns populations of squirrels or other mammals. Freshwater fish are inundated by saltwater.

The list of catastrophic effects is long, but the end result is that the local populations suffer and must rebuild until the next hurricane comes along. Other catastrophic events may include climatic changes or the accidental dumping of toxic waste. Oil spills have become an increasing threat to coastal populations of all types. In a truly severe event the populations may never return and the members that survive may not be able to live in the wasted environment.

## Human Populations

Human populations defy the ecological rules imposed on other animal populations. Because we can modify our environment, humans can live beyond the carrying capacity of our environment by growing extra food and building shelter. Medicine has helped the survivorship of our young and elderly. It is hard to describe human populations in terms of density dependence or independence. We are a highly successful species that is increasingly intelligent about survival in formerly inhospitable habitats.

Most scientists agree, however, that there is a limit to which the human population can grow. Eventually all the food that can be produced may still be inadequate to support the population. Resources like water and energy may reach their limits. There is no reason to believe that human populations will not also be regulated by environment. Mass starvation has already occurred in regions where natural catastrophes destroy food resources and hinder the attempts of other populations to help. There are scientists and governments that advocate population regulation on a voluntary basis to

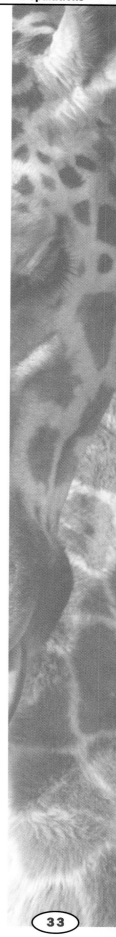

keep the human population from exceeding the carrying capacity of the Earth.

## Extinction

Extinction is a part of population biology. It is a natural process that has been or will be experienced by every population. The fossil record is full of animals that flourished for millions of years and then vanished.

Extinction and the mechanisms that compose it are not always well understood. The Earth is still experiencing extinctions at an amazing rate. The mammoths and saber-toothed cats are among the better known populations to have vanished in the last ten to twenty thousand years. Animal populations still disappear every day. However, when one population disappears, quite often another grows. Humans are blamed for the loss of many species. As human populations stabilize, additional populations will begin to flourish. This is the nature of population ecology cycles. SEE ALSO HUMAN POPULATIONS; POPULATION DYNAMICS.

*Brook Ellen Hall*

### Bibliography

Gutierrez, A. P. *Applied Population Ecology: A Supply-Demand Approach*. New York: John Wiley & Sons, 1996.

MacArthur, Robert H., and Joseph H. Connell. *The Biology of Populations*. New York: Wiley, 1996.

# Porifera

The phylum Porifera contains all the species of sponges. Phylogenetically, Porifera is most closely related to Protista, making it the first animal phylum to have evolved to be multicellular. This also makes Porifera the simplest in form and function. Sponges arose 550 million years ago in the pre-Cambrian period, evolving from colonial protists, groups of identical single cell organisms that live together. Evidence for this comes from specialized cells called choanocytes which sponges use in feeding. Although sponges are made up of many cells with specialized functions, their cells are not organized into true tissues. This lack of true tissue layers makes sponges different from all other animals except protozoans, which are not multicellular. Sponges also lack symmetry, true organs, a digestive or respiratory system, a nervous system, muscles, and a true mouth.

**sessile** not mobile, attached

Sponges are **sessile**; they are attached to one place and do not move around. They range in size from over 1 meter (3 feet) long to 2 millimeters (less than 1/8 of an inch) long. All sponges live in water, from the deepest seas to the shallow coastal waters. Most species are marine and can be found in all the oceans; only 3 percent live in fresh water. All sponges have the ability to completely regenerate an adult from fragments or even single cells. Sponges reproduce sexually, with one sponge producing both sperm and eggs from the choanocytes at different times, giving rise to a larvae that is free living (not sessile). A very few species reproduce asexually by **budding**. Some of the first **naturalists** like Aristotle mistakenly thought sponges were plants because they do not move and can regenerate.

**budding** a type of asexual reproduction where the offspring grow off the parent

**naturalists** scientists who study nature and the relationships among the organisms

Sponges depend on the water currents flowing through them for food and gas exchange. Sponges have specialized cells for gathering small particles of food from the water and distributing the food around the organism. Water comes in through pores along the body wall into the **spongocoel**, the main cavity of a sponge, and flows out a large opening in the top called an osculum. Choanocytes, also called collar cells, are specialized feeding cells which line the spongocoel. Choanocytes have a **flagellum** that extends out of the cell and sweeps food particles into a sticky, collarlike opening. They are similar in shape and function to certain colonial protists, such as the choanoflagellates. Amoebocytes, which digest food and transport it around the sponge, are specialized cells that move around the sponge's body under the **epidermis**, the outer layer of cells, through a jellylike middle cell layer. Amoebocytes move in a way that is similar to how amoebae move. Amoebocytes secrete hard structural fibers called spicules, which are made of calcium carbonate or silica. In some sponges, amoebocytes secrete other materials that make up the skeleton called spongin which are flexible fibers made of collagen. Only sponges have spicules. This structural feature is part of what divides sponges into different classes.

There are over nine thousand identified species of sponges, and more are identified all the time. These species are classified into three classes: Demospongiae, Calcarea, and Hexactinellida.

Most species of sponges are in the class Demospongiae. Sponges in this class are mostly marine, but the class also contains the few species that do live in fresh water. Because the materials that make up the skeleton and spicules of these sponges are so varied; the overall sizes and shapes of the sponges are also varied. The amoebocytes of the sponges in Demospongiae contain pigment, giving these sponges many different bright colors.

Sponges within the class Calcarea are characterized by spicules made of calcium carbonate. All species in Calcarea have spicules of a similar size and shape. Most species are not colored. Calcarea sponges are usually less than 15 centimeters (6 inches) tall, and live in the shallow ocean waters along coasts.

Glass sponges make up the class Hexactinellida. They are unique because their spicules have six points and a hexagon shape. The spicules fuse together to form elaborate lattice skeletons which make the sponges look as if they are made of glass. Most Hexactinellida live in the Antarctic Ocean and are found in deep waters, from 200 meters (650 feet) down. SEE ALSO PHYLOGENETIC RELATIONSHIPS OF MAJOR GROUPS.

*Laura A. Higgins*

**spongocoel** the central cavity in a sponge

**flagellum** a cellular tail that allows cells to move

**epidermis** the protective element of the outer portion of the skin found in some animals; it is composed of two layers of cells where the outer layer is continuously shed and replaced by the inner layer

**Bibliography**

Anderson, D. T., ed. *Invertebrate Zoology.* Oxford: Oxford University Press, 1998.

Barnes, Robert D. *Invertebrate Zoology,* 5th ed. New York: Saunders College Publishing, 1987.

Campbell, Neil A., Jane B. Reece, and Lawrence G. Mitchell. *Biology,* 5th ed. Menlo Park, CA: Addison Wesley Longman, Inc., 1999.

Purves, William K., Gordon H. Orians, H. Craig Heller, and David Sadava. *Life: The Science of Biology,* 5th ed. Sunderland, MA: Sinauer Associates Inc. Publishers, 1998.

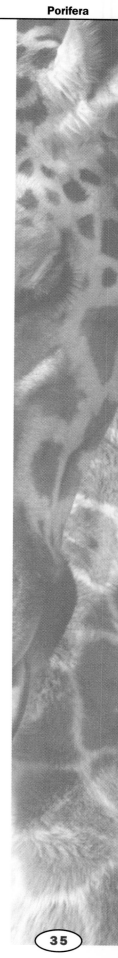

# Predation

Predation is the interaction in which the predator attacks live prey and consumes it. The interaction can be between two or more individuals, and is to the benefit of the predator and at the expense of the prey. The study of predator-prey interactions is broad and includes behaviors of the predator (such as searching, handling, and consuming prey), adaptations of the prey (survival strategies), and phenomena of their coexistence, such as stabilization factors that allow both groups to persist. It should be noted that there are four types of predators: true predators (including cannibals), grazers, parasitoids, and parasites. This entry will focus on true predators.

If a predator were fully efficient, all of its prey would be eaten. The prey would go extinct, and so would the predator. But the predator-prey interactions seen in nature allow both to sustain themselves. The first researchers to model how these interactions operated were A. J. Lotka and V. Volterra, in 1925 and 1926 respectively.

The Lotka-Volterra model assumes that predator reproduction is a function of the amount of prey consumed, so that when predators eat more prey, the predators increase in number through increased reproduction and immigration. There is a circular pattern of predator-prey interactions in this model: (1) when the predator population increases, the prey population decreases; (2) when the prey decrease in number, the predators decrease in number; (3) when the predator population decreases, the prey population increases; and (4) when the prey increase in number, the predator population again increases and the cycle begins anew.

When populations are plotted out over time, a pattern of coupled oscillation can be seen in which the apex, or peak, of one population coincides with the low point of the other. The numeral values of the two populations then cross and reverse positions.

A famous example of coupled oscillation between predator and prey populations occurs with the snowshoe hare and the lynx. Population peaks, as determined by numbers of pelts lodged with the Hudson Bay Company, were alternately spaced in time, with that of the lynx closely following that of the showshoe hare. The Lotka-Volterra model easily explains this pattern of predator-prey population sizes.

Although this model is not incorrect, it does oversimplify the scope of predator-prey interactions given that the major assumption, that when predators eat more prey the predator population increases, is often not exactly what is seen in nature. In actuality, when prey increases, a predator can have a numerical response, where predators do in fact increase in number through reproduction or immigration, or a functional response, where each predator eats more prey items.

Three types of functional responses are recognized, each of them showing a different relationship between prey density and amount of prey consumed. The Type I functional response is a direct relationship in which the predator eats all of the prey available up to a certain saturation point, when the predator can eat no more. After the predator reaches that saturation point, the prey density can continue to increase with no effect on how many prey items are being eaten. Some insects employ the strategy of

Many large predators, such as mountain lions, tigers, and wolves, are on the federal list of endangered or threatened species. One of the reasons is that these animals need large areas to hunt; another reason is habitat loss.

A grizzly bear finds a red salmon lunch in Brooks Falls, Katmai National Park, Alaska.

having thousands of offspring that all hatch at once. This suddenly floods the food supply, ensuring that a significant portion will remain after predators eat their fill.

The Type II functional response is more commonly seen because it is more realistic, since it incorporates a factor called handling time. Handling time is the amount of time a predator must devote to each prey item it consumes. It is the time needed for pursuing, subduing, and consuming the prey, and then preparing for further search. In this type of response the relationship between prey density and consumption is not linear because it changes over time. At first, the consumption rate increases, but as prey density continues to increase, there is a decline in the rate at which consumption increases until a maximum level is reached. This gradual deceleration of consumption reflects the factor of handling time.

Lastly, the Type III functional response is the most complex. It is similar to Type II at high prey densities, but includes the additional factor that there is very little or no prey consumption when prey is at low densities. This means that the predator does not eat any of the prey until there is a certain amount available.

One reason for this is that when there are very few prey animals, they can all find ideal hiding places and easily keep themselves out of the reach of predators. When there are more prey, however, some are forced into less ideal refuges, or into foraging places that are out in the open, where they are more visible to predators.

Another reason that prey are often not consumed when they are at low densities relates to search images. A predator gets accustomed to looking in certain types of habitat for certain shapes, colors, or movement patterns in order to hunt at maximum efficiently. Using a search image for the prey items that are most abundant pays off, because the predator will have the most success in hunting that prey. Searching for something that is very rare,

Population cycles in snowshoe hare and lynx.

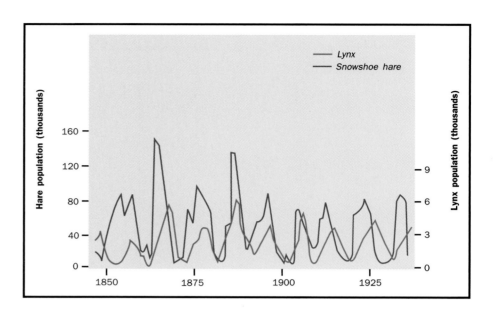

on the other hand, only wastes time and likely results in less food obtained in a given amount of time.

Related to the idea of search image is the phenomenon of switching. Even though a predator may have a preference for one type of prey, at times when that prey is at low densities and other prey is at high densities, the predator will switch to an alternate prey that is at a high density.

All three of these factors—the ability of prey to hide, a search image for the predator, and prey switching by the predator—combine to result in little or no prey taken when prey densities are particularly low. This allows prey populations to recover. Then, the predators increase their consumption until handling time again becomes a limiting factor. When this happens, the rate of consumption increase slows down and consumption evens out at a maximum.

A cannibal is a special type of predator. The term "cannibal" is applied to an individual that consumes another individual of the same species. Typically, cannibalism appears when there is simply not enough food available; in dense populations that are stressed by overcrowding (even when food is adequate); when an individual is weakened and vulnerable to attack as a consequence of social rank; and when vulnerable individuals, such as eggs and nestlings, are available. Frequently, the larger individuals do the cannibalizing, which can serve the purposes of obtaining a meal and reducing competition for food, mates, or territory in the future. SEE ALSO FOOD WEB; FORAGING STRATEGIES; INTERSPECIES INTERACTIONS.

*Jean K. Krejca*

**Bibliography**

Begon, Michael, John L. Harper, and Colin R. Townsend. *Ecology*, 2nd ed. Cambridge, MA: Blackwell Scientific Publications, 1990.

Lotka, A. J. *Elements of Physical Biology*. Baltimore, MD: Williams & Wilkins, 1925.

Ricklefs, Robert E., and Gary L. Miller. *Ecology*, 4th ed. New York: W. H. Freeman, 2000.

Smith, Robert L. *Elements of Ecology*, 2nd ed. New York: Harper & Row, 1986.

# Primates

Primates are one group (order) of mammals that evolved about 65 million years ago (mya). The early primates were probably small tree-dwelling (*arboreal*) animals that hunted for insects at night. Today we recognize at least 167 living species in the order primates. We can classify these species (and the primate fossils we find) into two major groups (suborders): prosimians (at least 38 species) and anthropoids (at least 129 species). Prosimians include lemurs (Lemuriformes), galagos and lorises (Lorisiformes) and tarsiers (Tarsiiformes). Anthropoids include New World monkeys (Platyrrhini; from South America), Old World monkeys (from Africa and Asia), apes and humans (Catarrhini). However, some scientists feel that tarsiers should be classified as anthropoids rather than as prosimians.

## Prosimians

The first primates to evolve were prosimians. These early primates split into several subgroups about 55 mya. One lineage (see graph) gave rise to Lorisiformes and Lemuriformes, and the other lineage gave rise to Tarsiiformes and the ancestors of all other primate groups (anthropoids). The first New World monkeys probably evolved about 45 mya, followed by Old world monkeys about 40 mya. The first apes appeared about 25 mya, human ancestors (e.g. *Australopithecus*) about 4 mya, and humans (*Homo*) about 2 mya.

Many modern prosimians still resemble the early primates in appearance and life style. For example galagos, lorises, tarsiers and some lemurs (e.g. from the dwarf lemur family) are all small **arboreal** animals that are active at night (nocturnal), hunting for insects or feeding on gum, tree sap, fruit, flowers or leaves. **Nocturnal** prosimians range in size from 60 g (mouse lemur) to about 2800 g (aye-aye). Only the larger lemurs (about 2–10 kg) have made the transition to being active during the day (diurnal). We assume that their increased size allowed them to make this transition because larger animals are less susceptible to being caught by **diurnal** predators. Diurnal lemurs mostly feed on fruits, flowers or leaves, and some of them may even come down from the trees to travel on the ground (e.g. ring-tailed lemurs and sifakas). The largest prosimians that ever lived were gorilla-sized lemurs on the island of Madagascar. However, when humans colonized the island about 1000 years ago these large lemurs were hunted to extinction. Lemuriformes occur exclusively on the island of Madagascar. Lorisiformes can be found all over Africa and Asia, and Tarsiiformes are restricted to Asia.

## Anthropoids

All anthropoids (monkeys, apes and humans) are diurnal with one exception: the owl monkey from South America. Monkeys occur in the Old and New World, but apes are restricted to Africa and Asia. Humans probably originated in Africa, but are now distributed worldwide. Monkeys and apes range in size from about 150g (pigmy marmosets) to about 180 kg (gorillas). They have variable diets including insects, fruits, leaves, meat and grains. Among the Old-World monkeys and apes we find many terrestrial species that have left the trees for the ground. One advantage of **terrestrial** life is that these primates can now use their hands more freely to manipulate objects rather than holding on to branches, and that they can gather in larger social groups.

**arboreal** living in trees

**nocturnal** active at night

**diural** active through the daytime

**terrestrial** living on land

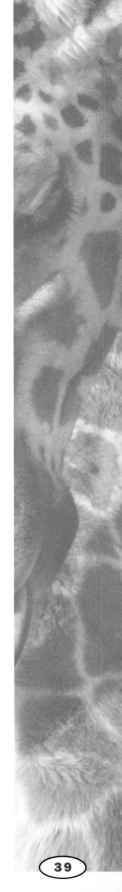

The front and hind paws of a Western lowland gorilla. The opposable thumbs visible in this image are defining characteristics of primates.

The social life of primates is much more complex than that of other mammals. Among primate species we find a wide range of gregariousness from the small nocturnal primates who spend most of their time looking for food alone, only occasionally meeting others (e.g. mouse lemurs), to primates living in small family groups (e.g. tamarins), and to those that live in large social groups (e.g. baboons) with many individuals of different ages, sex, dominance ranks and relatedness. Individuals within such large groups may compete with each other for food and mating partners, they may collaborate together against others, and may even form friendships.

## Primate Evolution

Many of the primate characteristics that we see in primates today (e.g. opposable thumbs, nails instead of claws, forward facing eyes) probably evolved as an adaptation to life in trees. For example primate hands have opposable thumbs that allows them to grasp tree branches, but also allows the handling of tools in apes and humans. The replacement of claws by nails went along with an increased sensitivity of the hands and fingertips and an improved ability to manipulate objects. During primate evolution we also see

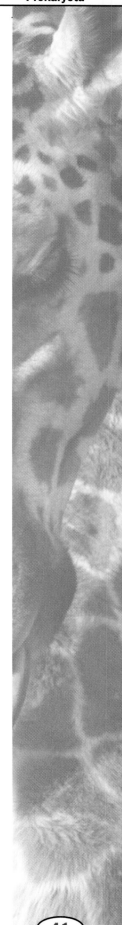

a trend from having eyes on the side of the face, as most lemurs do, to having both eyes up front as in monkeys, apes and humans. Moving the eyes up front reduces the visual field on the side of the head, but it increases the overlap of the area seen by both eyes simultaneously, allowing for improved depth perception. Now distances (e.g. between branches) can be better estimated, and vision overall is improved. As the eyes are moved up front we also see a shortening of the snout and a reduction of the sense of smell. As a consequence visual signals are now replacing chemical signals as the main form of social communication.

## Dispersal

There are still many open questions about primate evolution and how primates became dispersed to the localities where we find them today. For example we do not know how lemurs reached Madagascar. When prosimians evolved in Africa about 65 mya, Madagascar had already separated from the mainland. One possibility that is being discussed is that a few prosimians may have reached Madagascar by rafting on large trees floating in the ocean, and upon arrival, gave rise to all the lemur species. As an alternative our estimated timeline of primate evolution may be incorrect and prosimians may have evolved much earlier. However, this is highly unlikely and not supported by fossil evidence. A similar problem is posed by the exclusive presence of New World monkeys on the South American continent, which was isolated from Africa and North America when anthropoids evolved. Rafting from North America or from Africa is being discussed, but morphological evidence and the direction of past ocean currents seems to favor Africa as the origin of New World monkeys. And finally, the discovery of new anthropoid fossils from Asia in the 1990s suggests that the first anthropoids may have evolved in Asia rather than in Africa.

In addition to the dispersal questions there are still many puzzling questions about the evolution of the diverse primate social systems. For example researchers are studying which factors may determine whether primates live alone or in groups, or which circumstances may shape the nature of the interactions between individual primates.

Anthropologists are interested in studying non-human primates in the hope to gain insight into the evolution of the complex human social behavior. At present, 50 percent of all non-human primate species are threatened by extinction, and unless their habitats are protected from the ever-expanding human population in the near future, many primates will be lost forever.

*Kathrin F. Stanger-Hall*

### Bibliography

Fleagle, John G. *Primate Adaptation and Evolution.* Academic Press, 1988.

Smuts, Barbara B., Dorothy L. Cheney, Robert M. Seyfarth, Richard W. Wrangham, and Thomas, T. Struhsaker, eds. *Primate Societies.* Chicago: The University of Chicago Press, 1987.

### Internet Resources

Beard, Chris. "Vertebrate Paleontology." *Carnegie Museum of Natural History.* <http://www.primate.wisc.edu/pin/factsheets/links.html>.

*Primate Ancestors in China.* <http://www.cruzio.com/~cscp/beard.html>.

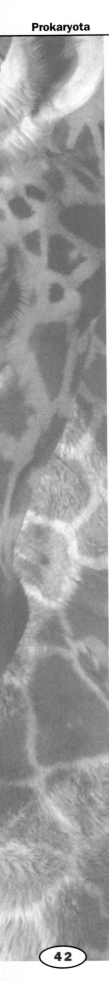

# Prokaryota

The prokaryota are one of the two major groups of biological organisms. The other is the **eukaryota**. Prokaryotes consist of two kingdoms: the archaebacteria and the eubacteria. Prokaryotes contribute the greatest **biomass** (amount of living matter) of any biological group and inhabit virtually all known earthly environments.

## Characteristics of Prokaryotic Cells

Prokaryotes exhibit a comparatively simple cellular organization compared to eukaryotes. For example, they lack membrane-bound **organelles** such as chloroplasts and mitochondria. They also lack a nucleus, and their DNA exists in the form of a single, circular chromosome that floats freely in the **cytoplasm**. Sometimes additional DNA is present in the cell in the form of smaller loops called plasmids. All prokaryotes are characterized by a cell membrane and a cell wall. Prokaryotic cell walls are composed of peptidoglycans, which are composed of amino acids and sugar. While some eukaryotes also possess cell walls, theirs are composed of different compounds.

Prokaryotes are sometimes described by their shape. Cocci are round, baccilli are rod-shaped, and spirochetes are helical. Prokaryotes are largely asexual and reproduce by fission (splitting up). However, some exchange of genetic material between individuals does occur via a process called conjugation.

The prokaryota are an extremely varied group. Some prokaryotes are heterotrophs (which obtain nutrients from other biological organisms), while others are **autotrophs** (which create their own nutrients from inorganic matter and energy sources). Some prokaryotic autotrophs photosynthesize. Others use resources not exploited by any known eukaryotes, such as hydrogen, ammonia, and compounds of sulfur or iron.

## Major Groups of Prokaryotes

Because of the comparative simplicity of the prokaryotes' morphology (shape and structure), much of the classification within the group relies on features of chemistry, metabolism, and physiology, in addition to shape, motility (ablity to move), and structural features.

**The archaebacteria.**   The archaebacteria, or **archaea**, are prokaryotes that inhabit some of the harshest environments that exist. They are extremely primitive, and retain some of the features of the earliest living cells. They differ from other prokaryotes, and from all living organisms, in the unusual lipids that are found in their cell membranes. Many aspects of their biochemistry are similarly distinctive.

There are three groups of archaebacteria. Species have been classified by their physiological characteristics and ecology(habitat, food resources used, etc.), rather than by their phylogenetic relationships (the sequence of branching events in evolutionary history which have resulted in the production of divergent species), which remain unclear. There is some evidence that archaebacteria may be more closely related to the eukaryotes than to the other group of prokaryotes, the eubacteria.

The methanogens are obligate anaerobes that can only survive in oxygen-free environments. Methanogens use hydrogen gas as an energy source. The

**eukaryota** a group of organisms containing a membrane bound-nucleus and membrane-bound organelles

**biomass** the dry weight of organic matter comprising a group of organisms in a particular habitat

**organelles** membrane bound structures found within a cell

**cytoplasm** a fluid in eukaryotes that surrounds the nucleus and organelles

**autotrophs** organisms that make their own food

**archae** an ancient linage of prokaryotes that live in extreme environments

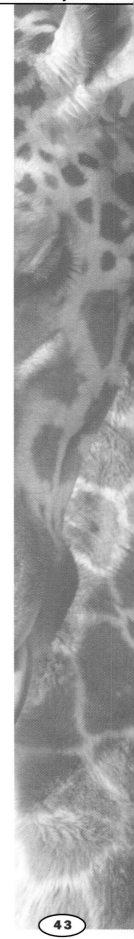

symbiotic bacteria that live in the guts of species such as cows and sheep and assist in the digestion of plant material are methanogens. They are so named because methane gas is a by-product of their metabolism.

The second group of archaebacteria, the extreme halophiles, inhabit very salty environments such as the Dead Sea and the Great Salt Lake. Finally, the extreme thermophiles, which represent several distinct lineages of archaebacteria, live in very hot environments, near hot springs or in deep ocean thermal vents. It is believed that the extreme thermophiles were the first living creatures on earth.

**The eubacteria.** The eubacteria form an extremely diverse group. Some major subgroups of eubacteria are discussed below.

The cyanobacteria are also known as blue-green algae. They perform **photosynthesis**, and many species can fix nitrogen (take in atmospheric nitrogen and incorporate it into the body of the organism) as well. Cyanobacteria are found in aerobic environments in which light and water are available.

**photosynthesis** process of converting sunlight to food

Some cyanobacteria engage in **mutualistic relationships** with fungi to form lichens. They also form a major component of oceanic plankton. Stromatolites are impressive chalk deposits that result from the binding of calcium-rich sediments by large colonies of cyanobacteria.

**mutualistic relationships** symbiotic relationships where both organisms benefit

Spirochetes are eubacteria with a unique spiral-shaped morphology. Spirochetes move using structures called undoflagella or axial filaments, which are similar to bacterial flagella but allow spirochetes to move by rotating the way a corkscrew rotates. Most species are free living, and some are **pathogens**; spirochetes are responsible for diseases such as syphilis and Lyme disease.

**pathogens** disease-causing agents such as bacteria, fungi, and viruses

The enterics are a group of rod-shaped eubacteria that live in the intestinal tracts of other organisms. This group includes the well-known *Escherichia coli* and its relatives. Enterics all ferment glucose. They are part of the normal gut **flora** of humans and have been extensively studied.

**flora** plants

The myxobacteria are significant for having the most complex life cycles among prokaryotes. Myxobacteria aggregate, or come together, to form multicellular "fruiting bodies" that give rise to spores. They live in the soil.

Vibrios are eubacteria characterized by a curved rod shape. They are found primarily in aquatic habitats. Cholera is caused by a vibrio.

Rickettsias and chlamydiae are two groups of obligate parasites or pathogens. Rickettsias require hosts in order to obtain nutrients, while chlamydiae obtain ATP (adenosine-triphosphate, the organic molecule which forms the basis of energy in all living organisms) from host cells. SEE ALSO EUKARYOTA; KINGDOMS OF LIFE.

*Jennifer Yeh*

**Bibliography**

Cano, Raul J. and Jaime S. Colome. *Microbiology.* St. Paul: West Publishing Co., 1986.

Curtis, Helena. *Biology.* New York: Worth Publishers, 1989.

Gould, James L., and William T. Keeton. *Biological Science*, 6th ed. New York: W. W. Norton and Co., 1996.

Singleton, Paul. *Bacteria in Biology, Biotechnology, and Medicine.* New York: J. Wiley and Sons, 1997.

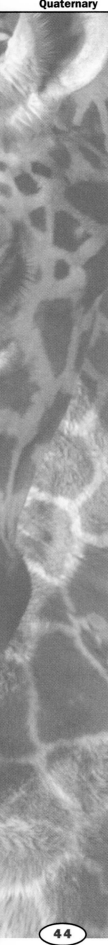

# Quaternary

The Cenozoic era, 65 million years ago to the present, is divided into two periods, the Tertiary and the Quaternary. The Tertiary period, 65 to 2 million years ago, encompasses the rebuilding of the animal kingdom at the end of the great Cretaceous extinction. From an unpromising beginning as small, nocturnal opportunists, mammals, along with the surviving birds, radiated into the most numerous and diverse life forms to inhabit Earth.

During the Quaternary, 2 million years ago to the present, dramatic climate changes reduced the overall diversity of animal life, yet saw the rise of another remarkable evolutionary opportunist—the **bipedal hominid**.

The earlier phase of the Quaternary, the Pleistocene epoch (2 million to 10,000 years ago), covers the alternating periods of glacial advance and retreat across the Northern Hemisphere and the corresponding effect on plant and animal life. The current chapter of Earth history, the Holocene epoch, deals with the interglacial period which saw the rise to dominance of *Homo sapiens*, the first mammal to shape its environment as well as be shaped by it.

During the Pleistocene, the continents continued their leisurely drift northward. Antarctica remained in place over the South Pole with a steadily increasing ice cap. The land bridge that formed between North and South America continued to disrupt ocean currents by sending colder water into the tropics. These geographic events, coupled with an overall cooling trend of about 10°F, seem to have tipped the planet into a series of "ice ages" interspersed with more brief, temperate periods. During glaciation, ice sheets formed at the North Pole moved across the Northern Hemisphere, locking up huge amounts of global water. Sea levels dropped by as much as 90 meters (300 feet), connecting previously separated land masses. Europe and Britain became contiguous, as did Siberia and Alaska. The more important effect of the glaciers was the decreasing of worldwide humidity. As the tropics became deserts, the larger mammals of Madagascar and Australia were pushed to extinction. As forests declined into grasslands, animals worldwide were pushed into shrinking refuges and competition with one another. Herds of herbivores that thrived on grasslands were unable to cope when the temperate forests expanded northward during the warmer periods. The grasslands in the colder latitudes closer to the pole were not as lush or varied and were nutritionally poorer. The episodic advance and retreat of habitat may have led to the contemporary patterns of migration in warm-blooded animals.

Continual climate stresses and the mingling of species across what had once been barriers contributed to overall decreased diversity. By the end of the Pleistocene 12,000 years ago, the "megafauna"—mammoths, mastodons, 2.5-metric-ton (3-ton) sloths, and *Aepyornis*, the 400-kilogram (900-pound) bird—had all but disappeared. Almost unnoticed among the giants (mammoths, mastodons, and sloths) of the time was a new evolutionary player. Arboreal primates resembling small tree shrews had flourished 60 million years ago giving rise to several groups. These were the **placental** mammals, which had hands with the ability to grasp, eyes that faced forward, and teeth adapted to an **omnivorous** diet. Hand-to-eye coordination, encouraged by

**bipedal** walking on two legs

**hominid** belonging to the family of primates

**placental** having a structure through which a fetus obtains nutrients and oxygen from its mother while in the uterus

**omnivorous** eating both plants and animals

| Era | Period | Epoch | Million Years Before Present |
|-----|--------|-------|------------------------------|
| Cenozoic | Quaternary | Holocene | 0.01 |
| | | Pleistocene | 1.6 |
| | Tertiary | Pliocene | 5.3 |
| | | Miocene | 24 |
| | | Oligocene | 37 |
| | | Eocene | 58 |
| | | Paleocene | 66 |

Quaternary period and surrounding time periods.

a tree-climbing lifestyle, also forced the development of the "thinking" part of the brain.

As environmental stresses worked on the prosimians, they developed into the brainier anthropoids, and finally hominoids, the apelike family that spread throughout Africa, Europe and Asia 25–10 million years ago. Around 4 million years ago, driven by need or lured by opportunity, certain primates took up a new behavior, walking on their hind legs. Bone fossils show that their originally upright posture alternated with four-legged running and climbing. But gradually, the hominid became more and more bipedal, freeing its hands for carrying and manipulating its environment. Upright walking required dramatic changes in anatomy, and these changes further widened the anatomical gap between this protohuman and its closest relative, the quadripedal ape.

Pelvic changes limited the size of the young at birth, creating a longer period of infant dependency. This in turn encouraged the development of a social organization to protect and rear the young. Other benefits of this cooperation fostered more complex arrangements such as foraging together, using of communal shelters, tool making, the specialization of labor, and sharing resources. Thus began the long journey from the tropical treetops into the remotest regions of the globe, by the mammal which possesses the power to affect all life on Earth.

The history of life can be seen as an unending cycle of environmental pressure → change and decline → **adaptive radiation** → and more pressure. *Homo sapiens*, the first species ever to create its own environmental pressures, now stands at the threshold. Will we adapt or decline? SEE ALSO GEOLOGICAL TIME SCALE.

*Nancy Weaver*

**adaptive radiation** a type of divergent evolution where an ancestral species can evolve into an array of species that are specialized to fit different niches

**Bibliography**

Asimov, Isaac. *Life and Time*. Garden City, NY: Doubleday & Company, 1978.

Fortey, Richard. *Fossils: The Key to the Past*. Cambridge, MA: Harvard University Press, 1991.

————. *Life: A Natural History of the First Four Billion Years of Life on Earth*. New York: Viking Press, 1998.

Friday, Adrian, and David S. Ingram, eds. *The Cambridge Encyclopedia of Life Sciences*. London: Cambridge University, 1985.

Gould, Stephen Jay, ed. *The Book of Life*. New York: W. W. Norton & Company, 1993.

Lambert, David. *The Field Guide to Prehistoric Life*. New York: Facts on File, 1985.

McLoughlan, John C. *Synapsida: A New Look Into the Origin of Mammals*. New York: Viking Press, 1980.

Steele, Rodney, and Anthony Harvey, eds. *The Encyclopedia of Prehistoric Life*. New York: McGraw Hill, 1979.

Wade, Nicholas, ed. *The Science Times Book of Fossils and Evolution*. New York: The Lyons Press, 1998.

# Reproduction, Asexual and Sexual

Organisms must reproduce and, in the context of evolution, must choose among different methods to do so. There are two major strategies for reproduction—sexual and asexual. Each tactic has its own advantages and disadvantages, and each is appropriate for certain situations. Vertebrates, such as humans, are almost exclusively sexual in their reproduction, many simpler animals are asexual. To decide which reproductive strategy may prove advantageous in a given set of circumstances, it is important to understand how they differ.

## Asexual reproduction

**Asexual reproduction** takes a variety of forms. The simplest one-celled organisms may reproduce by binary fission, in which the cells simply divide in half. This form of reproduction creates a clone of the parent, and has the benefit of usually being very quick and energy efficient. For example, bacteria that reproduce by binary fission can give rise to **progeny** every few hours.

Some organisms, such as *Cryptosporidium parvum*, a **sporozoan** that causes traveler's diarrhea, may utilize multiple fission, in which they split into more than one offspring simultaneously. In multicellular organisms, a similar tactic is called fragmentation. In this process, small pieces break off and grow into new organisms. Still other organisms reproduce by **budding**, in which a smaller copy of the parent grows on the body and eventually splits off to begin life on its own.

All these variations of asexual reproduction have one thing in common, the offspring is a direct clone of the parent. The purpose of reproduction is to propagate one's own genes. Evolutionarily, asexual reproduction is a good deal for the parent. It is quick, simple, and the genes of the parent will not be diluted by those of another individual. In addition, an organism that reproduces asexually can reproduce about twice as fast as one that reproduces sexually. This has shown to be true with the whiptail lizard of the southwestern United States, which can reproduce both sexually and asexually under different conditions.

## Sexual reproduction

Sexual reproduction is more much complex than asexual reproduction. It requires the production of sex cells, or **gametes**, which have half the number of **chromosomes** of all other cells in the organism. When the organism needs to make sex cells, it undergoes meiosis, which produces **haploid cells** (one copy of the **genome**) from **diploid cells** (two copies of the genome). A key aspect of meiosis is that the two copies of a single chro-

---

**asexual reproduction** a reproduction method with only one parent, resulting in offspring that are genetically identical to the parent

**progeny** offspring

**sporozoans** parasitic protozoans

**budding** a type of asexual reproduction where the offspring grow off the parent

**gametes** reproductive cells that only have one set of chromosomes

**chromosomes** structures in the cell that carry genetic information

**haploid cells** cells with only one set of chromosomes

**genome** an organism's genetic material

**diploid cells** cells with two sets of chromosomes

---

There are a few species of vertebrates that reproduce asexually. The whiptail lizard, which lives in the desert grasslands of the southwestern United States, may reproduce sexually or asexually. Do asexually-reproducing lizards show less genetic variability than sexually-reproducing ones? They do, just like the theory says they should.

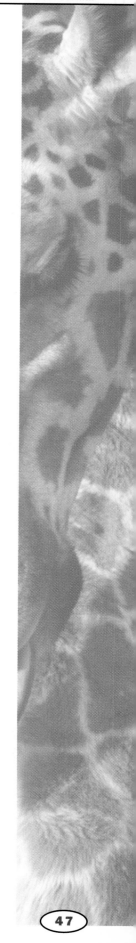

mosome can cross over to create a completely new chromosome that contains a new combination of genes. The net effect of crossing-over is that genes on a specific chromosome can change position from one chromosome to the next. This means that genes from both parents may end up next to each other on the same chromosome. Where genes are concerned, switching from chromosome to chromosome is a good way to ensure they will keep active in a given population.

Once the gametes are made in the male and female, they must meet with one another to form offspring. The sperm from the male provides one copy of a genome. The egg from the female provides another copy of a different genome. Thus, the offspring of sexually reproducing organisms has more than one opportunity to switch genes around—crossing-over and the union of the two parents.

## Comparing Sexual and Asexual Reproduction

However, note how much energy sexual reproduction takes. The sex cells must be made, and as each parent contributes only half the genome, it propagates only half as many genes from each offspring as does an asexually reproducing organism. Recall that an organism is most interested in propagating its genes; indeed, that is the whole point of reproduction. To reproduce sexually is to reduce the amount of genetic material one reproduces by half, and this reduction does not even take into account the effort sexually reproducing organisms must make to find mates, then impress, select, or defend them. Nevertheless, nearly all higher animals reproduce sexually. Why? The answer to this question is far from settled, but biologists have a few good clues.

The most important thing about sexual reproduction is its ability to switch around successful genes. If it is beneficial to an organism's survival to be both tall and have blue eyes, a short, blue-eyed parent and a tall, brown-eyed parent can get together and stand a good chance of producing offspring with both characteristics. If they reproduced asexually, a short, blue-eyed parent would have to wait around for a height-inducing genetic mutation to change height and eye color. And because mutations, which are basically genetic mistakes, tend to cause bad effects, the mutation rate in most organisms is exceedingly slow. While it would take only a generation for sexually-reproducing parents to beget tall offspring with blue eyes, it might take an asexually-reproducing parent hundreds or thousands of generations!

Asexually reproducing organisms do not readily share genetic material, but they do reproduce much faster. And because asexually reproducing organisms reproduce faster, they do exceptionally well in situations where they have no competition. With sexually-reproducing competition nearby, however, the asexual organisms will quickly be outadapted and outevolved by their neighbors, even though the asexual organisms may have superior numbers due to fast reproduction. Many biologists think that intense competition gives rise to sexual reproduction, because the competition requires rapid innovation and distribution of the most successful genes.

Although these arguments for the existence of sexual reproduction might seem evolutionarily sound, the alleged advantages of sexual reproduction

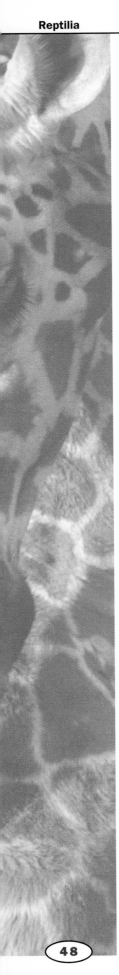

over asexual reproduction are still quite controversial among biologists. Some biologists think that only replicating half of your genes in exchange for sexual reproduction is not an even trade. Others suggest that dilution of groups of genes does not matter. Furthermore, a sexually reproducing organism must expend a great amount of effort to find a mate, in both behavior and new body structures and appendages. Biologists believe that **sexual selection** drives gender size and appearance, plumage, behavior, and many other energetically expensive strategies.

**sexual selection** selection based on secondary sex characteristics that leads to greater sexual dimorphism or differences between the sexes

Can it be possible that sexual selection, with all its demands, is worth the moderate amount of recombination that results from sexual reproduction? If not, why do all vertebrates, many invertebrates, and most plants sexually reproduce? Many prominent biologists have considered these questions, such Richard Dawkins, J. Maynard Smith, G. C. Williams, and others.

It seems likely that the ability to swap around already successful genes, rather than being forced to sit around and waiting for mutations, is a more successful strategy for complex organisms. And less complex organisms can get by without the larger energy and resource investment that sexual reproduction demands. SEE ALSO EGG; EMBRYONIC DEVELOPMENT; FERTILIZATION; REPRODUCTIVE SYSTEM.

*Ian Quigley*

**Bibliography**

Conn, David Bruce. *Atlas of Invertebrate Reproduction and Development.* New York: John Wiley and Sons, 2000.

Curtis, Helena, and N. Sue Barnes. *Biology,* 5th ed. New York: Worth Publishing, 1989.

Hayssen, Virginia, Ari Van Tienhoven, Ans Van Tienhoven, and Sydney Arthur Asdell. *Asdell's Patterns of Mammalian Reproduction: A Compendium of Species-Specific Data.* Ithaca, NY: Comstock Publication Associates, 1993.

Norris, David O., and Richard E. Jones, eds. *Hormones and Reproduction in Fishes, Amphibians, and Reptiles.* New York: Plenum Press, 1987.

Purves, William K., and Gordon H. Orians. *Life: The Science of Biology.* Sunderland, MA: Sinauer Associates Inc., 1987.

## Reproductive System   *See Endocrine and Reproductive System.*

# Reptilia

Most reptiles can be classified into three large groups: the turtles (order Chelonia), the snakes and lizards (order Squamata), and the alligators and crocodiles (order Crocodilia). Most reptiles share a number of general morphological features. In general, reptiles are lung-breathing vertebrates with two pairs of limbs and a horny, scaly skin. Reptiles are **amniotes**, which means that their large, yolky eggs have a protective layer called an **amnion** which prevents them from drying out on land. Rather than laying eggs, some snakes and lizards bear their young live.

**amniotes** vertebrates which have a fluid-filled sac that surrounds the embryo

**amnion** the membrane that forms a sac around an embryo

**poikilothermic** an animal that cannot regulate its internal temperature; also called cold-blooded

Nonavian reptiles are **poikilothermic** (cold-blooded) creatures, which means that they derive their body heat from external sources (in contrast to homothermic animals that maintain a constant body temperature through internal mechanisms). Contrary to popular belief, the "cold-bloodedness"

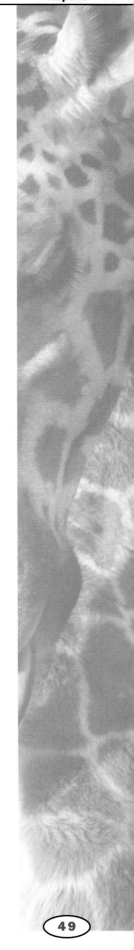

of reptiles does not mean that they maintain low body temperatures. Reptiles control their body temperature through a process called thermoregulation, and their internal temperature can fluctuate greatly according to their surroundings. Researchers have found that many reptiles exert precise control over body temperature by moving around to different areas within their surrounding habitat.

## Chelonia

Turtles are classified into two sister groups, Pleurodira and *Cryptodira*. Pleurodires (side-necked turtles) bring their head and neck against their body by bending the neck to the side. Most cryptodires fully retract the head and neck into the shell, although certain groups such as sea turtles and snapping turtles have lost this ability. All pleurodires are aquatic, while cryptodires include **terrestrial**, aquatic, and marine forms.

**terrestrial** living on land

Turtles are easily distinguished from other reptiles by the protective shell that encases their body. The turtle shell has two parts, a carapace on top and a plastron under the belly, which incorporate the ribs, vertebrae, and elements of the turtle's **pectoral** (front limb) girdle. The turtle's head, limbs, and tail can protrude from and retract into the shell to varying degrees, depending on the species. The plastron of the box turtle has a hinge that allows the shell to snap shut and provide further protection for the head.

**pectoral** of, in, or on the chest

Turtles can have many different forms of shells. The bony elements of the shell can be covered with scaly, armored plates, (all land turtles and most aquatic and marine turtles) or leathery skin (soft-shelled turtles and the ocean-dwelling leatherback turtle). Terrestrial turtle shells have relatively high domes; the shells of water-dwelling turtles are flatter and more streamlined. Most extant turtles have shells that are less than 60 centimeters (2 feet) long, although sea turtles can grow up to a shell length of 2.7 meters (9 feet).

All turtles lay eggs. They do not guard their eggs or young while they are developing, but instead leave the eggs buried in loose dirt or sand. The survival rate of eggs and newborn turtles is very low, since turtles will produce large numbers of offspring to compensate for high mortality rather than fewer, stronger offspring that would have greater potential to survive. Some turtles lay multiple groups of eggs (called clutches) during each season, and the clutches of sea turtles may contain up to two hundred eggs. Developing embryos get their nutrients from the yolk in the egg and only become **carnivorous** or omnivorous after they hatch. Most turtles have a varied diet of small insects and worms, crustaceans, nonfibrous plants, or fish. In many turtle species, the sex of the offspring is determined by the temperature at which the eggs are maintained as the young turtles develop.

**carnivorous** describes animals that eat other animals

## Crocodylia

Crocodilians are large, amphibious, and carnivorous reptiles. Their long bodies, powerful jaws, and muscular tail are covered with heavy armor with bony plates. These aggressive predators have conical teeth and short legs with webbed, clawed toes. The largest crocodilian, the Nile crocodile, reaches a length of about 6 meters (20 feet). Many scientists believe that crocodilians are the closest living relatives to birds, but this relationship is still under debate.

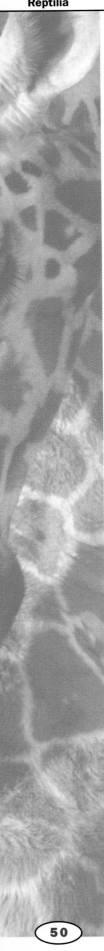

There are three major groups in Crocodilya—alligators, crocodiles, and gavials. These three groups are distinguished by characteristics of the teeth and jaw, although all crocodilians are similar with respect to ecology, morphology, and behavior.

Crocodilians are found in swamps, rivers, and lakes of tropical and subtropical habitats in the Northern and Southern Hemispheres. They spend most of their time in the water but are also known to travel long distances over land. They move from place to place in several ways, including sculling, in water (propelling themselves using a side-to-side tale motion), belly sliding (into water), high walking, and galloping. Crocodilians are sit-and-wait predators who float passively in the water while awaiting potential prey. The can knock large mammals into the water with a swing of their powerful tail. With their powerful jaws, some species clamp onto their victim's legs and tear the prey apart by rotating themselves rapidly in the water in an "alligator roll."

Crocodilians lay eggs in nests. After laying her eggs, the mother keeps them warm by covering them with mud and decaying plant material. She guards her eggs until the squeaks of the young signal that the babies are ready to hatch. The mother then knocks the dirt off of the eggs to help the young hatch, but provides no additional care for her young.

## Rhynchocephalia

**extant** still living

Two **extant** species of tuatara (genus *Sphenodon*) belong to this sister group of the squamates. The tuatara is lizardlike in general appearance, but is distinguishable from lizards by a number of characters involving tooth type and arrangement, skull morphology, and genitalia. The tuatara is the only nonavian reptile without a organ used exculsively for sexual intercourse. Like birds, male tuataras transfer sperm to the female while their cloaca (an opening through which feces is expelled, and through which reproduction occurs) are pressed together.

Tuataras are found only in New Zealand, where they live in burrows they construct or those that have been abandoned by other animals. During the daytime they bask in the sun near the entrance to their burrows, but are most active at night. They feed on insects, bird eggs, and young birds. Both species of tuatara are egg layers, and newborn young emerge after a thirteen-month incubation period. Tuataras reach sexual maturation after about twenty years and can live up to the age of fifty. Adult tuataras may reach a total length of up to almost 60 centimeters (two feet).

The human introduction of certain animals to New Zealand has placed many tuatara populations under threat of extinction. Tuataras are slow-moving, and are thus easily captured and eaten by rats, cats, and pigs. An estimated 100,000 tuataras remain on offshore islands near New Zealand. There are no tuataras on the New Zealand mainland. About half of the suriving tuatara live on Stephens Island in Cook Strait, and there have been proposals to relocate some individuals to sanctuaries on other islands so that more of them may be seen in the wild. Tuataras are currently listed as an endangered species by conservation groups, and human interaction with this species is strictly regulated.

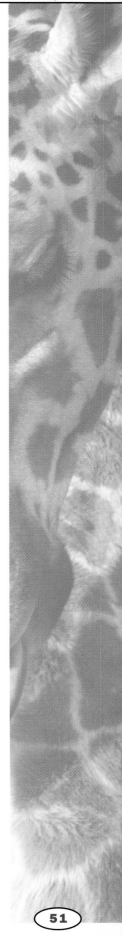

# Squamata

The squamates are the most diverse group of non-avian reptiles. Squamata consist of lizards (Sauria), snakes (Serpentes) and amphisbaenians (Amphisbaenia). Lizards are divided into two major groups, Iguania and Scleroglossa. These two groups are generally distinguishable by differences in tongue morphology and function. Iguanians have fleshy tongues that are used while capturing prey. The mucous coating on the tongue helps the lizard to pick up small insects, vertebrates, and plant matter, and aids in swallowing. The tongue of the chameleon is a well-known example of this iguanian trait.

Scleroglossans have thin, forked tongues that are used in chemosensation ( the process of gathering information about their environment through detecting chemical cues). These lizards use their tongues for "smelling" the air around them and use their jaws for grabbing prey. Snakes and amphisbaenians are both scleroglossans. Iguanians are territorial sit-and-wait predators; scleroglossans are active foragers that generally do not guard territories.

Most of the three thousand species of saurians (lizards, except for snakes and amphisbaenians) have legs, movable eyelids, external ear openings, and long tails. Some species are limbless. Saurians range in size from the 3-centimeter (1.2-inch) long gecko to the 3-meter (10-foot) long Komodo dragon. All saurians are terrestrial, but some species may live in trees and other foliage, under rocks, or in burrows.

Saurians have different ways of reproducing. Most species are **oviparous** (egg-laying), while other species are **viviparous** (bearing live young). Among viviparous lizards, some species rely exclusively on the yolk of the egg to provide nutrients to the developing embryo. In these lizards, the eggs that are retained in the mother's uterus during embryonic development resemble the eggs of oviparous lizards but do not have a hard, **calcified** shell. Instead, there is a thin shell membrane through which water, gas (oxygen and carbon dioxide), and waste materials can pass. In a smaller number of species, the mother transfers nutrients to her young through a **placenta**. The placenta is a region of the mother's uterus characterized by a high density of blood vessels. Through these blood vessels, water, gases, and waste products pass between the blood of the mother and her developing young.

Most saurians are diurnal (active only during the daytime) and regulate their body temperature by basking in the sunlight during the early morning hours and hiding in the shade during hotter periods. Geckos have special physiological characteristics that allow them to pursue a nocturnal (active during the nighttime) lifestyle. For example, the muscles of geckos are able to function under cooler temperatures than other lizards are use to. Saurians spend much of their time feeding, mostly on insects. Larger species also feed on other lizards and on small mammals. Several iguanian species feed exclusively on plants.

Serpents (snakes) are secretive, solitary predators that are found in almost all types of habitat worldwide, except near the North and South Poles. They are limbless, lack eyelids, and have only one lung. Their slim, long body form allows them to search for prey in burrows, nests, and crevices. Many snakes are also excellent tree climbers and swimmers, and several species are fully aquatic. Like saurians, snakes as a group exhibit both egg-laying and live-bearing forms of reproduction.

**oviparous** having off-spring that hatch from eggs external to the body

**viviparous** having young born alive after being nourished by a placenta between the mother and offspring

**calcified** made hard through deposition of calcium salts

**placenta** a structure through which a fetus obtains nutrients and oxygen from its mother while in the uterus

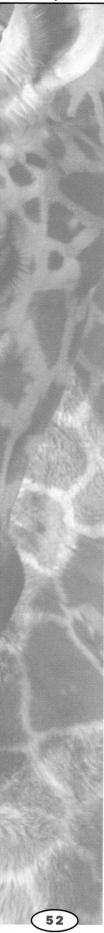

THe underside of the backfoot of a Tokay Gecko.

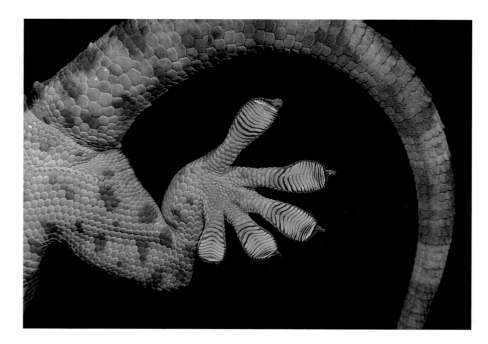

**infrared** an invisible part of the electromagnetic spectrum where the wavelengths are shorter than red, heat is carried on infrared waves

Snakes rely mainly on their senses of sight and smell for information about their surroundings. Individuals use chemical signals to communicate to each other, especially during courtship. Some species of boids (such as boa constrictors) and vipers (such as asps and adders) have also evolved heat-sensing **infrared** receptors called pit organs. Studies have shown that many snake species that lack pit organs can also sense infrared radiation, although to a lesser degree.

Amphisbaenians are elongate, burrowing squamates. With the exception of the genus *Bipes*, which has forelimbs, amphisbaenians are limbless. All species lay eggs. They occupy habitats ranging from tropical rain forests to deserts. SEE ALSO PHYLOGENETIC RELATIONSHIPS OF MAJOR GROUPS.

*Judy P. Sheen*

**Bibliography**

Grasse, Pierre-Paul, ed. *Encyclopedia of the Animal World: Reptiles.* New York: LaRousse & Co., 1975.

Pough, F. H., Robin M. Andrews, John E. Cadle, Martha L. Crump, Alan H. Savitzky, and Kentwood D. Wells. *Herpetology.* Upper Saddle River, NJ: Prentice Hall, 1998.

Stebbins, Robert C. *A Field Guide to Western Reptiles and Amphibians.* 2nd ed. Boston: Houghton Mifflin, 1985.

# Respiration

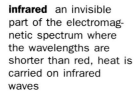

**adenosine triphoshate** an energy-storing molecule that releases energy when one of the phosphate bonds is broken; often referred to as ATP

**terrestrial** living on land

Organisms respire in order to obtain oxygen and get rid of carbon dioxide. Oxygen drives cellular metabolism, the process in which glucose from food is converted to usable energy resources in the form of **adenosine triphosphate**, (ATP). Carbon dioxide is a by-product of this reaction. Aquatic species obtain oxygen from water, while **terrestrial** species obtain it from air. These two media, water and air, are often associated with different respiratory strategies. This is partly because the amount of oxygen in air is far greater than that dissolved in water. In addition, air requires much less energy to pump than does water, which is considerably more dense.

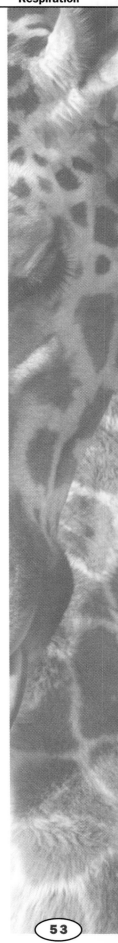

Many organisms do not require special respiratory organs because they obtain an adequate supply of oxygen through diffusion across the body surface. This is known as cutaneous gas exchange. Cutaneous gas exchange is often employed by small animals, which have a high ratio of surface area to volume. Some larger animals, such as certain annelid worms, are also able to use cutaneous exchange because of their large surface areas and modest energy demands.

Larger organisms almost invariably possess special respiratory organs. Respiratory specializations can be grouped into three major categories: **gills**, lungs, and **trachea**. All three types of organs evolved to provide extensive surface areas for use in gas exchange.

## Gills

Gills describe respiratory structures formed from external extensions of the body. They characterize diverse species, including some annelids, some arthropods such as **horseshoe crabs** and crustaceans, mollusks, certain **echinoderms**, and vertebrates such as fish and larval amphibians. Gills are typically found in aquatic environments, although some terrestrial species, such as terrestrial crustaceans, make use of them as well. Species with very active lifestyles tend to have more highly developed gills, with considerable surface areas for gas exchange. In the fishes, for example, a series of bony **gill arches** supports many primary **gill filaments**, each of which has numerous minute folds referred to as the secondary gill lamellae.

Oxygen is absorbed into the bloodstream as oxygenated water flows over the gills. Some aquatic species rely on natural currents to carry oxygenated water to them. Others expend energy to force water over the gills. Among fishes, ram ventilators swim rapidly with open mouths, forcing water to flow into the mouth, across the gills, and out through the spiracle or **operculum**. Active swimmers, such as sharks and tuna, often employ ram ventilation.

Other fish species use a complex series of mouth and opercular expansions and compressions to pump water over the gills. Many aquatic species, including fishes, are characterized by countercurrent exchange, a system in which oxygenated water and deoxygenated blood flow in opposite directions, maximizing the amount of oxygen **absorption** that occurs.

## Lungs

Lungs characterize many terrestrial species. Unlike gills, lungs are series of internal branchings that function in respiration. In humans, for example, air flows initially into the trachea. The trachea then splits into two bronchi, which split several more times into smaller and smaller bronchioles. These end at small sacs called **alveoli**, which are closely surrounded by capillary blood vessels.

It is in the alveoli that gas exchange actually occurs, with oxygen diffusing across the alveoli into the bloodstream. The alveoli are lined with substances called surfactants, which help to prevent them from collapsing. Humans have approximately 150 million alveoli. As with gills, the surface area available for gas exchange can be quite large.

Lungs characterize a diverse array of terrestrial species, including gastropod mollusks such as snails and slugs, spiders (whose respiratory organs

**gills** site of gas exchange between the blood of aquatic animals such as fish and the water

**trachea** the tube in air-breathing vertebrates that extends from the larynx to the bronchi

**horseshoe crabs** a "living" fossil in the class of arthropods

**echinoderms** sea animals with radial symmetry such as starfish, sea urchins, and sea cucumbers

**gill arches** arches of cartilage that support the gills of fishes and some amphibians

**gill filaments** the site of gas exchange in aquatic animals such as fish and some amphibians

**operculum** a flap covering an opening

**absorption** the movement of water and nutrients

**alveoli** thin-walled sacs in the lungs where blood in capillaries and air in the lungs exchange gases

are called book lungs because they contain a series of lamellae that resemble pages), and most terrestrial vertebrates.

Gas exchange using lungs depends on ventilation, the process through which air is brought from the external environment into the lungs. Ventilation mechanisms vary from taxon to taxon. In amphibian species, air is actively gulped, or forced into the lung by positive pressure. Reptiles, birds, and mammals use negative pressure to ventilate the lungs. These species expand the volume of the **thoracic** cavity where the lungs lie, causing air to be drawn into the lungs.

**thoracic** related to the chest area

Creating negative pressure in the thoracic cavity is accomplished in a variety of ways. Lizards and snakes use special muscles to expand their rib cages. Turtles extend forelimbs and hindlimbs out of their shells to create negative pressure. In crocodiles, the liver is pulled posteriorly, towards the rear of the animal, in order to expand the thoracic cavity. Mammals employ a combination of contracting their diaphragm, a muscular sheet that lies at the base of the thoracic cavity, and expanding the rib cage.

Birds have evolved unusually efficient respiratory systems, most likely because of the tremendous energy required for flight. The bird respiratory system includes large, well-developed lungs and a series of air sacs connected to the lungs and trachea. Some of the air sacs occupy the hollow spaces in larger bones such as the humerus and femur. The air sacs are involved in ventilating the lungs, and allow for a one-directional flow of air through the respiratory system. This is unique among the terrestrial vertebrates.

The respiratory exchange system of birds is described as crosscurrent exchange. Crosscurrent exchange is less efficient than the countercurrent exchange system of fishes. However, because the concentration of oxygen in air is much greater than in water, this method is extremely effective.

Cutaneous respiration may supplement gas exchange through lungs or gills in many species. It is critical in most amphibians, which are characterized by moist skins consisting of live cells in which oxygen can dissolve and then be taken up and used by the organism. Some amphibians have evolved special adaptations that make gas exchange across the skin more effective. In some aquatic **salamanders**, for example, the skin has become highly folded, an adaptation that increases the surface area available for gas absorption. There is even a large family of salamanders, the family Plethodontidae, in which the lungs have been lost entirely. This group relies almost entirely on cutaneous exchange. Similar developments are seen in other groups, including some species of slugs.

**salamander** a four-legged amphibian with an elongated body

## The Tracheal System

The tracheal system is a respiratory system that is unique to air-breathing arthropods such as millipedes, centipedes, and insects. The tracheal system consists of air-filled tubes that extend into the body from pores on the body surface known as spiracles. These tubules provide oxygen to tissues directly. Centipedes have one spiracle per segment, and millipedes have two per segment. The number of spiracles in insects varies, but there can be as many as ten pairs.

The spiracle is merely an opening to the environment in primitive insect species. However, more advanced groups can close their spiracles, and

The body of a caterpillar magnified to show its spiracles.

some are even outfitted with filtering devices. The ability to close the spiracle is advantageous because it allows for water conservation. Interestingly, the limits of the tracheal gas exchange system are believed to restrict insects to their generally small size. Insects, unlike animals such as vertebrates, do not use their circulatory systems to aid in the transport of oxygen.

Water is lost from all respiratory surfaces in air-breathing organisms. This is particularly serious for species that rely on cutaneous exchange, and explains why most species of amphibians are limited to fairly moist habitats.

Respiration is regulated either by the central nervous system or by more localized mechanisms. Vertebrates have a respiratory pacemaker in the medulla of the brain. In addition, respiratory rates are influenced by internal gas concentrations. Aquatic species usually regulate respiration based on internal oxygen levels, whereas terrestrial species tend to rely on internal carbon dioxide levels. SEE ALSO DIGESTION.

*Jennifer Yeh*

**Bibliography**

Curtis, Helena. *Biology*. New York: Worth Publishers, 1989.

Gould, James L., and William T. Keeton. *Biological Science*, 6th ed. New York: W. W. Norton, 1996.

Hickman, Cleveland P., Larry S. Roberts, and Allan Larson. *Animal Diversity*. Dubuque, IA: William C. Brown, 1994.

Hildebrand, Milton, and Viola Hildebrand. *Analysis of Vertebrate Structure*. New York: John Wiley, 1994.

Withers, Philip C. *Comparative Animal Physiology*. Fort Worth, TX: Saunders College Publishing, 1992.

# Respiratory System

All animals require oxygen. Oxygen enables living things to metabolize (or burn) nutrients, which releases the energy they need to grow, reproduce, and maintain life processes. Some animals can exist for months on fats or other foods stored in their bodies, and many can live a shorter time without

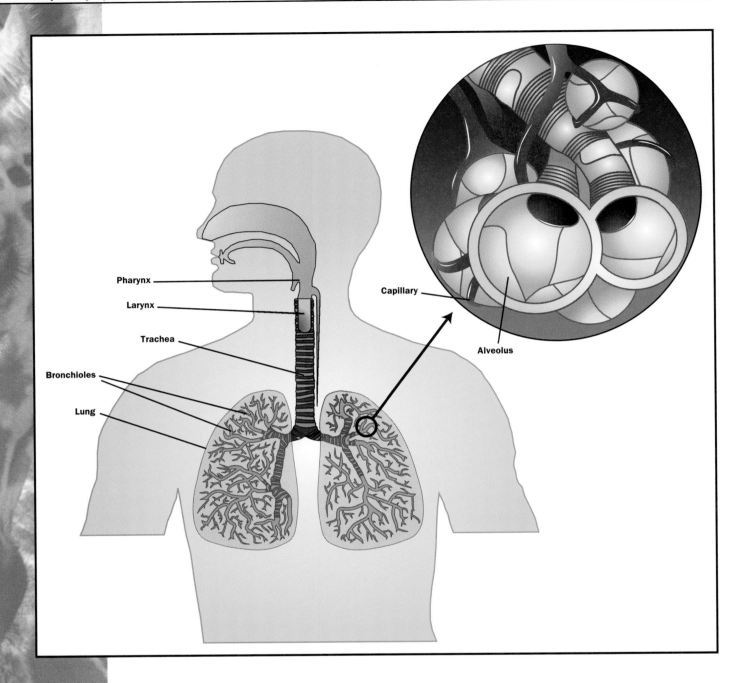

Pharynx

Larynx

Trachea

Bronchioles

Lung

Capillary

Alveolus

The human respiratory system. Redrawn from Johnson, 1998.

water. Yet few can survive for long without oxygen because little can be stored in the body. Most animals obtain oxygen from their environment. It is generally believed that life originated in the oceans, where many animals still live, obtaining their oxygen in a dissolved form from the water. In the course of evolution, various animals have become earth dwellers and have developed structures that allow them to breathe air.

Along with supplying oxygen, the respiratory system assists in the removal of carbon dioxide, preventing a dangerous and potentially lethal buildup of this waste product. The respiratory system also helps regulate the balance of acid and bases in tissues, which is a crucial process that enables cells to function. Without the prompt of conscious thought, the respiratory system carries out this life-sustaining activity. If any of the functions of the system are interrupted for more than a few minutes, serious and irreversible damage to body tissues would occur, and possibly result in death.

## Respiratory Systems of Various Species

Depending on the animal, the organs and structures of the respiratory system vary in composition and complexity. The respiratory system is composed of the organs that deliver oxygen to the circulatory system for movement or transport to all of the cells in the body. Organs or systems such as body covering, gills, lungs, or trachea allow the movement of gases between the animal and its environment. These structures vary in appearance but function in a similar way by allowing gases to be exchanged.

Animals obtain oxygen in a number of ways: (1) from water or air through a moist surface directly into the body (protozoan); (2) from air or water through the skin to blood vessels (earthworms); (3) from air through gills to a system of air ducts or trachae (insects); (4) from water through moist gill surfaces to blood vessels (fishes, amphibians); (5) from air through moist lung surfaces to blood vessels (reptiles, mammals, humans).

In one-celled **aquatic** animals, such as protozoans, and in sponges, jellyfish, and other aquatic organisms that are a few cell layers thick, oxygen and carbon dioxide diffuse directly between the water and the cell. This process of diffusion works because all cells of the animal are within a few millimeters of an oxygen source.

Insects, centipedes, millipedes, and some arachnids have fine tubes or trachea connecting all parts of the body to small openings on the surface of the animal. Movements of the thoracic and abdominal parts and the animal's small size enable oxygen and carbon dioxide to be transported from the trachea to the blood by way of diffusion.

In more complex animals, respiration requires a blood circulatory system and gills, in combination with blood, blood vessels, and a heart. Many aquatic animals have gills, thin-walled filaments that increase the surface area and increase the amount of available oxygen. The oxygen and carbon dioxide exchange occurs between the surrounding water and the blood within the gills. The gills of some larvae and worms are simply exposed to the water, while some aquatic crustaceans, such as crayfish, have special adaptations to force water over their gills. The gills of fishes and tadpoles are located in chambers at the sides of the throat, with water taken into the mouth and forced out over gills.

All land vertebrates, including most amphibians, all reptiles, birds and mammals have lungs that enable these animals to get oxygen from air. A heart and a closed circulatory system work with the lungs to deliver oxygen and to remove carbon dioxide from the cells. A lung is a chamber lined with moist cells that have an abundance of blood capillaries. These membranes take different forms. In amphibians and reptiles they can form a single balloon-like sac. In animals that require large amounts of oxygen, the lungs are a spongy mass composed of millions of tiny air sacs called **alveoli** that supply an enormous surface area for the transfer of gases.

In birds a special adaptation allows for the high-energy demands of flight. The lungs have two openings, one for taking in oxygen-filled air, the

**aquatic** living in water

**alveoli** thin-walled sacs in the lungs where blood in capillaries and air in the lungs exchange gases

other for expelling the carbon dioxide. Air flows through them rather than in and out as in the other lunged vertebrates.

## Respiratory System in Humans

The human respiratory system consists of the nasal cavity, throat (pharynx), vocal area (larynx), windpipe (trachea), bronchi, and lungs. Air is taken in through the mouth and or nose. The nasal passages are covered with mucous membranes that have tiny hairlike projections called cilia. They keep dust and foreign particles from reaching the lungs.

Approximately halfway down the chest the trachea or windpipe branches into two bronchi, one to each lung. Each branch enters a lung, where it divides into increasingly smaller branches known as bronchioles. Each bronchiole joins a cluster of tiny airsacs called alveoli. The pair of human lungs contain nearly 300 million of these clusters and together can hold nearly four quarts of air. After oxygen has crossed the alveolar membrane, oxygen is delivered to the cells by the pigment **hemoglobin**, found in blood.

The lungs in humans are cone-shaped and are located inside the thorax or chest, in the cavity framed by the rib cage. One lung is on either side of the heart. The right lung has three lobes; the left has two lobes. A thin membrane known as pleura covers the lungs, which are porous and spongy. The base of each lung rests on the diaphragm, a strong sheet of muscle that separates the chest and abdominal cavities.

The respiratory center at the base of the brain is a cluster of nerve cells that control breathing by sending impulses to the nerves in the spinal cord. These signals stimulate the diaphragm and muscles between the ribs for automatic inhalation. During inhalation the rib muscles elevate the ribs and the diaphragm moves downward, increasing the chest cavity. Air pressure in the lungs is reduced, and air flows into them. During the exhalation, the rib muscles and diaphragm relax and the chest cavity contracts. The average adult takes about sixteen breaths per minute while awake and about six to eight per minute while sleeping. If breathing stops for any reason, death soon follows, unless breathing movements are artificially restored by mouth to mouth breathing. SEE ALSO CIRCULATORY SYSTEM; TRANSPORT.

*Leslie Hutchinson*

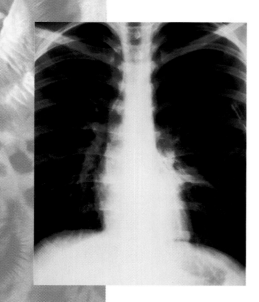

An x ray of the human lungs. The finely detailed structure of the lungs is difficult to see in this image.

**hemoglobin** an iron containing protein found in red blood cells that binds with oxygen

### Bibliography

Hickman, Cleveland, Larry Roberts, and Frances Hickman. *Integrated Principles of Zoology*, 8th ed. St. Louis, MO: Times Mirror/Mosby College Publishing, 1990.

Johnson, George B. *Biology: Visualizing Life*. New York: Holt, Rinehart and Winston, Inc., 1998.

Randall, David, Warren Burggren, and Kathleen French. *Eckert Animal Physiology: Mechanisms and Adaptations*, 4th ed. New York: W. H. Freeman, 1997.

### Internet Resources

"Respiratory System." *Britannica Online*. 1994–2000. Encyclopedia Britannica. <http://www.eb.com/>.

**Roundworms**  *See Nematoda.*

# Rotifera

The 1,500 to 2,000 species in the phylum Rotifera, like other members of the kingdom Animalia, are multicellular, heterotrophic (dependent on other organisms for nutrients), and lack cell walls. But rotifers possess a unique combination of traits that distinguish them from other animals, including **bilateral symmetry** and a **pseudocoelom**, a fluid-filled body cavity between two different layers of embryonic tissue. The pseudocoelom serves as a sort of circulatory system and provides space for a complete digestive tract and organs. It is also found in the closely related phylum Nematoda, a very common group of roundworm species. Unlike nematodes, which tend to live in moist soil, rotifers generally inhabit fresh water, although some species can be found in salt water or wet soil. Rotifers (Latin for "wheel bearers") have an unusual mode of transportation: a crown of beating cilia surrounds their mouth, propelling the animal head-first through the water. Rotifers eat protists, bits of vegetation, and microscopic animals (such as young larvae), which they suck into their mouths with the vortex generated by their cilia. Their jaws are hard and their pharynx is muscular, allowing them to grind up their food.

Another unusual rotifer trait is parthenogenesis, an asexual mode of reproduction in which females produce unfertilized eggs which are essentially clones of the mother. Some parthenogenetic species are completely female, while others produce degenerate, or poorly developed, males whose sole purpose is to fertilize eggs under stressful environmental conditions. Fertilized eggs are more resistant to drying out than unfertilized eggs. When conditions are favorable again, females return to producing unfertilized eggs. Because parthenogenesis is rare, rotifers provide valuable clues to biologists trying to understand the evolution of sexual reproduction, which remains one of the great unsolved mysteries of evolution.

Most rotifers are between 0.1 to 0.5 millimeters (0.004 to 0.02 inches) long, and generally contain only a few hundred cells. Another unusual characteristic of the phylum is that the number of cell divisions during development is fixed. This means that the animal is unable to grow new cells in response to damage. One might view this inflexibility as a drawback, but the fact that most of the rotifer's closest relatives do not share this trait suggests that the rotifer's recent ancestors were capable of **regeneration**. This means that rotifers could have inherited the ability to regenerate, but did not. Perhaps the ability to regenerate has costs as well as benefits, and the costs outweighed the benefits for the ancestor of the rotifers. Such costs are that regeneration requires energy, cell division, and can result in cancer-causing mutations.

Both **asexual reproduction** and deterministic cell division are examples of traits that might be considered "primitive" because they are simpler or tend to appear in older taxa. Yet, by examining a phylum like Rotifera and understanding its evolutionary history, we realize that there are situations in which apparently "primitive" traits replace "advanced" ones. A complex

**bilateral symmetry** characteristic of an animal that can be separated into two identical mirror image halves

**pseudocoelom** a body cavity that is not entirely surrounded by mesoderm

**regeneration** regrowing body parts that are lost due to injury

**asexual reproduction** a reproduction method with only one parent, resulting in offspring that are genetically identical to the parent

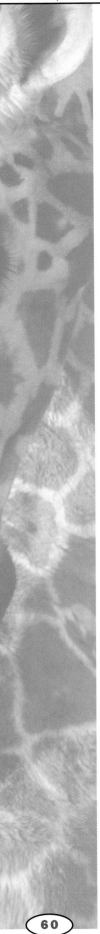

A freshwater rotifer with an embryo inside.

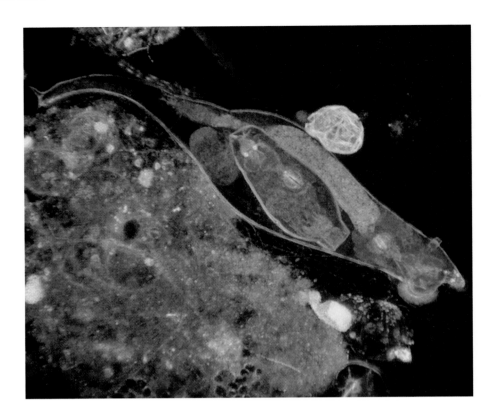

trait is not necessarily better than a simpler one. Each has its place in the diversity of life. SEE ALSO PHYLOGENETIC RELATIONSHIPS OF MAJOR GROUPS.

*Brian West*

**Bibliography**

Campbell, Neil A. *Biology*, 2nd ed. Redwood City, CA: The Benjamin/Cummings Publishing Company, Inc., 1990.

Curtis, Helena, and N. Sue Barnes. *Biology*, 5th ed. New York: Worth Publishers, 1989.

**integument** a natural outer covering

**exoskeleton** a hard outer protective covering common in invertebrates such as insects

# Scales, Feathers, and Hair

The term "**integument**" is applied to any outer covering of an animal. Basically, it means the skin, although many scientists describe the **exoskeleton** of arthropods as an integument. An exoskeleton is a coating of hard protein type substances that entierly cover the outside of the animal. It provides a place for muscle attachment. The vertebrate skeleton is internal and muscles attach from the outside. This permits larger growth of the animal. Vertebrate animals have developed some very interesting excrescences, or projections of the skin, some of which are well-known and easily oberved by most people. They are scales, feathers, and hair. Although these characteristics are common among the most vertebrates, they are not all that common in the animal world as a whole.

The primary function of the excrescences is protection and insulation. Claws, hooves, and nails are other types of excrescences, but do not provide the animal with the complete body protection it gets from the three main body coverings. All, however, are made from special proteins. The primary protein used by vertebrates is keratin. It occurs as two forms: alpha (more pliable) and beta (more stiff) keratin. The way in which the keratin is constructed at the molecular level is what determines the structural differences among the various excrescences.

## Scales

Scales occur on fishes and reptiles. Fish scales are made from more than just keratin. They are often derived from bone in the deeper layers of the skin (the middle section of tissues called the mesoderm), which are named the dermal layer. Some have a **placoid scale** that is formed from bone and covered with enamel, the same covering as the outer portion of a human tooth. Such scales are rough, spiny projections that give the surface of the fish a sandpaper-like feel.

**placoid scale** a scale composed of three layers and a pulp cavity

A primitive type of scale is the **ganoid scale**. While this type of scale was common to fishes that lived hundreds of millions of years ago, today it is primarily found in a fish named the gar. Scales of this type are similar to placoid scales, but have an extra coating of a very strong substance called ganoin. It is even stronger than enamel and these gar-type scales are very common in the fossil record. They are believed to have provided a great deal of protection against rocks and abrasive surfaces that fishes may have rubbed against. It is also possible that they provided some protection against predators, since the scales make it very hard to bite through the skin of fishes. Many scientists believe that the ganoid scale is the precursor or original type scale that led to tooth evolution.

**ganoid scale** hard, bony, and enamel covered scales

The **ctenoid scale** is found in most modern fishes and is much lighter than the placoid or ganoid scale. Growth rings are an interesting feature of a ctenoid scale. Under a microscope, small circular folds are apparent. There is some debate about the timing of the growth rings. Some researchers once believed they were annual rings, but this has been disproved. The growth rings appear with various timings among varieties of fishes and there are still many questions about what the growth rings may indicate about the life of fishes.

**ctenoid scale** a scale with projections on the edge like the teeth on a comb

In both fishes and reptiles, the scales cover the entire body. The construction of reptile scales, however, is different. Both lizards and snakes have scales over their entire bodies. Many species are identified by the pattern the scales make on the head and body. Reptiles have evolved a unique layer in the skin from their **aquatic** ancestors. In order for reptiles to live free from water a waxy layer, the stratum corneum, evolved to keep the animal from drying out on land. These waxy layers lie between the layers that produce the keratin for the scales. The scales are overlapping layers of skin with an inner and outer coating. A unique adaptation is a basal hinge region from which the scales can be somewhat flexible and fold back without falling off.

**aquatic** living in water

The stratum corneum, or cornified layer, contains beta keratin and in some species both beta and alpha keratin. This provides strength to the scale, which enables reptiles to climb on rocks and slither along various surfaces such as gravel and sand. As the animal grows, every so often **epidermis** (the outer covering) is sloughed off during a process called shedding. A new epidermis forms and is, in turn, shed as the animal continues to grow.

**epidermis** the protective portion of the outer portion of the skin found in some animals, it is composed of two layers of cells where the outer layer is continuously shed and replaced by the inner layer

This photomicrograph shows the shaft, barbs and barbules of a hummingbird feather.

**contour feathers** feathers that cover a bird's body and give shape to the wings or tail

**pterylae** feather tracks

There is a wide variety of scale structures among reptile groups, including turtles and crocodiles who are sometimes excluded by researchers as true reptiles. Colors are achieved by a wide variety of pigments that are incorporated into the scale or that lie in layers of the skin.

## Feathers

Feathers are interesting excrescences whose origins are not completely understood. It was once believed that only birds had feathers, but later research revealed a striking relationship between dinosaurs and birds. This debate has been recently reduced with the finding of new and important fossils in China. Scientists now see a much closer relationship, if not actual descent, between birds and dinosaurs. Dinosaur fossils discovered in China near the end of the twentieth century revealed a skeleton which is clearly that of a dinosaur but which has impressions of feathers all over the body. The most fascinating discoveries began in the mid 1990s and is ongoing. It is easy to observe the relationship between birds and dinosaurs and their contemporary reptile cousins by looking at a bird. A bird has scales over its legs and have many characters, including skull and skeleton, that are very reptilian. They are highly modified scales composed of beta keratin.

A feather is a scale in which a long center shaft, the rachis, is the dominant feature. On either side of the shaft, the keratin is divided into tiny barbs that, under a microscope, look like the close-knit leaves of a fern. The barbs have tiny hooks, or hamuli, on them, which help the barbs attach to one another and keep them close to each other. These interlocked barbs are called the vane of a feather.

There are three basic types of feathers. The **contour feathers** (tail and flight feathers) are long and used as an aerodynamic device for flight. The plumules (down feathers) are for insulation to keep the bird warm. Their barbs or barbules (smaller versions of barbs) are not closely knit and so the plumules appear fluffy. The feather is built somewhat like a fern frond. There are primary barbs that come of a sturdy shaft. The smaller barbules emerge from the barbs and help hold the feather in shape by interlocking the barbs. This design is crucial for keeping the feather sturdy during flight. The hair feathers (filoplumes) are not as fluffy as the down, but are still used for insulation.

Bird feathers are composed of beta keratin. They occur in tracts along the bird's skin called **pterylae**. These tracts are easily seen on plucked chickens or turkeys used for food. It surprises many people to learn that the entire bird is not covered with feathers.

## Hair

Hair is one of the most familiar excrescences, since humans are endowed with varying amounts of it. It is a characteristic of mammals. Hair or fur is made from beta keratin and grows from follicles located all over the epidermis. Its primary function is to provide insulation for the animal to keep it warm. However, there are several kinds of hair that provide sensory functions. Whiskers, or vibrissae, are located at places near a mammal's head. The roots of the vibrissae are connected to sensory nerves that are sensitive to movement and so help the animal to detect its environment.

The structure of hair is different from that of scales and feathers. A hair is basically a cone of keratin that is derived from keratinized cells in the dermis, or middle layers of skin. The hair is generated and formed in a pit in the skin called the follicle. The hair has an inner and outer sheath. Near the base of each hair attached to the follicle is a small muscle, called an erector pilli. When stimulated, this muscle contracts and pulls the hair straight up. In humans this condition is called goose bumps. They come as result of the tightening of the skin which helps prevent heat loss. It serves as a warning gesture in cats and dogs and other mammals.

As in most animals with scales and feathers, the loss of hair is cyclic. The heavy winter coats of many animals are shed in spring and replaced by lighter summer ones. Shedding may occur again in fall when a warmer coat is needed.

Whatever the excrescence, the variety and color of scales, feathers, and hair make watching animals a pleasant activity. The variety of patterns and structures continually provide a wonderful display of life on earth. SEE ALSO KERATIN.

*Brook Ellen Hall*

**Bibliography**

Hildebrand, Milton. *Analysis of Vertebrate Structure.* New York: John Wiley, 1998.

Moyle, Peter, and Joseph Cech Jr. *Fishes: An Introduction to Ichthyology.* Englewood Cliffs, NJ: Prentice Hall, 1992.

Sloan, Christopher. *Feathered Dinosaurs.* Washington, D.C.: National Geographic Society, 1999.

**Internet Resources**

*Encyclopedia Britannica.* <http://members.eb.com>.

*Hair: Animal Diversity Page.* Museum of Zoology at the University of Michigan. <http://animaldiversity.ummz.umich.edu/anat/hair.html>.

# Scientific Illustrator

How do toads unfurl their tongues? What goes on inside a car's engine? How do hummingbirds hover? What happens when atomic particles collide? Scientific illustrators have the fascinating job of turning such ideas and theories into clever colorful charts, diagrams, and three-dimensional models. Artists with a keen eye for detail and a penchant for science and technology work with engineers, scientists, and doctors to make information that is difficult to convey with words comprehensible at a glance. Their assignments may range from illustrating the spores of a fungus to depicting the surface of Mars or building a model of an atom to making a model of a prosthetic limb. Scientific illustrators work with computer graphics, plastics, and modeling clay as well as traditional techniques such as charcoal, pens, and brushes.

Some scientific illustrators are self taught. One of the most famous was the English naturalist and author, Beatrix Potter. Potter never attended school, yet spent her entire life studying the world around her in finer and finer detail, rendering her observations—especially those of fungi—in exquisite drawings which were presented to the Royal Academy of Sciences.

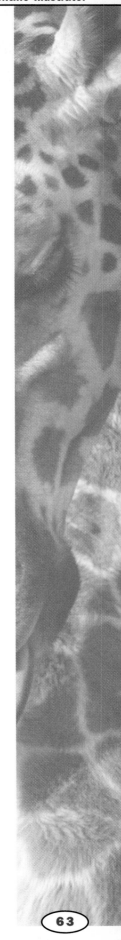

Students who wish to become scientific illustrators must have a strong background in the graphic fine arts, with an additional emphasis on biology, engineering, architecture, or design. Some illustrators work for specific magazines or publishing houses, but most are freelance and work from home. To be successful, scientific illustrators must be able to run a small business.

The field of medical illustration is more specialized, concentrating on human anatomy and physiology. Medical illustrators produce drawings for textbooks and models for anatomical displays, and may also be called upon to present visual testimony in court cases. To become certified, medical illustrators must complete a special master's program which is currently offered only to a handful of students each year at six schools in North America. Prerequisites for entrance to these programs are a bachelor's degree with majors and minors in art and biological science. The coursework includes vertebrate anatomy, embryology, physiology, chemistry, and histology.

*Nancy Weaver*

**Bibliography**

U.S. Dept. of Labor. *Occupational Outlook Handbook.* Washington D.C.: Bureau of Labor Statistics, 2000.

## Segmented Worms  *See Annilida.*

# Selective Breeding

Selective breeding is evolution by human selection. As nineteenth-century British naturalist Charles Darwin noted in *Variation of Animals and Plants under Domestication*, selective breeding may be methodical or unconscious. Methodical selection is oriented toward a predetermined standard, whereas unconscious selection is the result of biases in the preservation of valuable individuals. Methodical selection requires great care in discriminating among organisms and is capable of rapid change in specific traits, such as milk production or silk color. Unconscious selection, more common in ancient times, resulted in grains and seeds such as wheat, barley, oats, peas, and beans, and in animal traits such as speed and intelligence.

## Historical Overview

Selective breeding began about 10,000 years ago, after the end of the last Ice Age. Hunter-gatherers began to keep flocks and herds and to cultivate cereals and other plants. This process of domestication was probably stimulated by a combination of human population pressure and environmental stress caused by a rapid change in climate. **Global warming** at the end of the Ice Age created drought in areas where rainfall had previously provided sufficient water, forcing people to congregate around reliable water sources. The increased population density favored the cultivation of plant and animal species for use during times when they were not naturally plentiful.

**global warming** a slow and steady increase in the global temperature

## Selective Breeding vs. Natural Selection

Like natural selection, selective breeding requires genetic variation on which to act. If the variation in a trait is strictly environmentally induced, then the

An Aberdeen angus breed is shown at the 1993 Royal Highland Show in Scotland. Professional breeders would have carefully chosen the parents of these show animals in order to increase their chances of winning "best of breed."

selected variants will not be inherited by the next generation. Selective breeding also requires controlled mating. Thus, animals that are social and easily manipulated, such as **bovids**, sheep, and dogs, were easier targets for selective breeding than territorial species, such as cats and other carnivores. Cultures without a strong concept of property rights, such as those of pre-Columbian South America, were less likely to domesticate species because of their difficulty segregating different breeds. A short generation time also facilitates selective breeding by speeding up the response to selection. For example, most plants with multiple breeds (races) are annuals or biennials.

Selective breeding differs fundamentally from natural selection in that it favors **alleles** (forms of a gene) that do not contribute favorably to survival in the wild. Such alleles are usually **recessive**, for otherwise they would not persist in wild populations. Selective breeding is essentially a process of increasing the frequency of rare, recessive alleles to the point where they usually appear in homozygous form. Once the wild-type alleles are eliminated from the population, the process of domestication has become irreversible and the domestic species has become dependent on humans for its survival.

There is abundant evidence of the effectiveness of selective breeding. In general, there is more genetic variation among breeds of the same species for valuable traits than for others. For example, tubers are diverse among potatoes, bulbs are diverse among onions, fruits are diverse among melons. The implication is that selective breeding for valuable traits has created the diversity.

## Domestication

The earliest archaeological evidence of selective breeding has been found in the Near East, where plants and animals were domesticated 10,000 years

**bovids** members of the family bovidae which are hoofed and horned ruminants such as cattle, sheep, goats and buffaloes

**alleles** two or more alternate forms of a gene

**recessive** hidden trait that is masked by a dominant trait

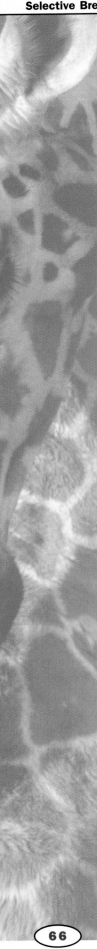

ago. China followed suit 2,000 years later, and sub-Saharan Africa, central Mexico, the central Andes, and eastern North America began selective breeding around 4,000 years ago. Historical evidence includes rules given for influencing sheep color in chapter thirty of Genesis, ancient Greek philosopher Plato's note that Glaucus selected dogs for the chase, Alexander the Great (356–323 B.C.E.) selecting Indian cattle, Roman poet Virgil's (70–19 B.C.E.) description of selecting the largest plant seeds, and the Roman emperor Charlemagne's selection of stallions in the ninth century. The Incas of Peru rounded up wild animals and selected the young and the strong for release, killing the rest. This strategy mimicked the action of natural selection, whereas elsewhere artificial selection used the most valuable individuals.

Selective breeding was invented independently in several different parts of the world, but its first appearance was in the Fertile Crescent, an **alluvial** plain between the Tigris and Euphrates Rivers. Ten thousand years ago, hunter-gatherers in a western part of the Crescent known as the Levantine Corridor began to cultivate three cereal crops: einkorn wheat, emmer wheat, and barley. Each was descended from a different wild species. A thousand years later, hunters in the eastern region of Zagros began to herd goats. Within 500 years after that, cereal cultivation and goat herding had spread to the center of the Crescent and combined with sheep and pig herding to form a diverse agricultural economy.

The process of domestication resembles a common mechanism of natural speciation. First, a barrier is created to separate a species into distinct reproductive groups within its geographical range. Over many generations the reproductively isolated groups begin to diverge as a result of selection, whether artificial or natural. All of a species' adaptations to artificial selection, both deliberate and incidental, are referred to as its "adaptive syndrome of domestication." Domesticated species eventually lose the ability to survive in the wild as part of their adaptive **syndromes**.

The path to domestication followed a stereotypical sequence of events. The first step in the domestication of a seed plant was the disturbance of the earth near settlements. This disturbed habitat facilitated the spread of pioneer plants that were adapted to natural disturbance. It also provided a colonization opportunity for the plants with seeds gathered, and dropped, by humans. The second step was the deliberate planting of seeds that were gathered from favored plants in the previous generation. Favored species tended to be pioneers adapted to growing in dense stands. One byproduct of this process was the selection of greater harvest yields. Since seeds that were collected were the only ones that reproduced, selection for increased seed production was strong. Another hallmark of domesticated plants was rapid sprouting, since competition between seedlings can be strong. The seeds themselves lost the ability to lie dormant and become larger. All of these traits were accidental byproducts of the storage and planting of seeds, rather than the results of methodical selection.

The domestication of animals also produced stereotypical traits. Animal species that were hardy, useful to humans, easy to breed in captivity, and friendly toward humans and each other tended to be successful targets for selective breeding. Of particular importance was flexibility of feeding habits, which facilitated human management. Solitary species with **idiosyncratic**

**alluvial** composed of sediments from flowing water such as silt, sand, mud, and gravel

**syndromes** groups of signs or symptoms that, occurring together, form a pattern

**idiosyncratic** a trait or manner peculiar to an organism

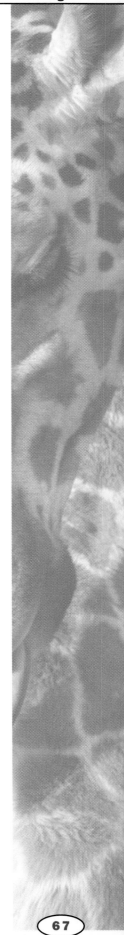

feeding behaviors were unlikely to reproduce successfully in captivity in spite of early agriculturalists' best efforts. Domestic animals were probably already an important source of food before they were domesticated. Just as plants became domesticated as a result of controlled reproduction, animals were reproductively isolated as herds and flocks. There is, however, only one hallmark of selective breeding in animals: small body size. The remainder of the adaptive syndrome of domestication is unique to each species. SEE ALSO DOMESTIC ANIMALS; EVOLUTION; FARMING; GENETICS; GENETIC VARIATION IN A POPULATION; NATURAL SELECTION.

*Brian R. West*

### Bibliography

Clutton-Brock, Juliet. *Domesticated Animals from Early Times*. Austin: University of Texas Press, 1981.

Darwin, Charles. *The Variation of Animals and Plants under Domestication*. Washington Square: New York University Press, 1988.

# Sense Organs

The nervous system is responsible for sensing the external and internal environments of an organism, and for inducing muscle movement. Human sensation is achieved through the stimulation of specialized neurons, organized into five different modalities—touch, balance, taste, smell, hearing, and vision. The touch modality includes pressure, vibration, temperature, pain, and itch. Some animals are also able to sense magnetism and electric fields. Modality, timing, intensity, and location of the stimulus are the four features that allow the brain to identify a unique sensation.

## Sensation Receptors

The neurons specialized to detect sensation are also called receptors because they are designed to receive information from the environment. Each receptor responds only to a stimulus that falls within a defined region, called its receptive field. The size of the stimulus can affect the number of receptors that respond, and the strength of the stimulus can affect how much they respond. For example, when a cat sits on your lap, a large population of receptors responds to the cat's weight, warmth, claws, and the vibrations from its purring.

Transduction refers to the transfer of environmental energy into a biological signal signifying that energy. Sensory receptors either transduce their respective stimuli directly, in the form of an **action potential**—the electrochemical communication between neurons—or they chemically communicate the transduction to a **neuron**. Neurons collect information at their **dendrites**, a long, branched process that grows out of the cell body. Information travels through the cell body, and then reaches the **axon**, another long process designed to transfer information by apposing, or synapsing on, another neuron's dendrites. In some receptors, the axon is branched, and is specialized for both initializing and transmitting action potentials.

**Touch receptors**.    Touch receptors are a type of mechanoreceptor because they are activated by mechanical perturbation of the cell membrane. The axon is located in either shallow or deep skin, and may be encapsulated by specialized membranes that amplify pressure. When the appropriate type of

**action potential** a rapid change in the electric charge of the cell membrane

**neuron** a nerve cell

**dendrites** branched extensions of a nerve cell that transmits impulses to the cell body

**axons** cytoplasmic extensions of a neuron that transmit impulses away from the cell body

This snake uses its tongue to smell the surrounding environment.

**dorsal** the back surface of an animal with bilateral symmetry

**proprioceptors** sense organ that receives signals from within the body

**polymodal** having many different modes or ways

**photoreceptors** specialized cells that detect the presence or absence of light

**retina** a layer of rods and cones that line the inner surface of the eye

pressure is applied to the skin, these membranes pinch the axon, causing it to fire. The action potential travels from the point of origin to the neuron's cell body, which is located in the **dorsal** root ganglion. From there, it continues through another branch of the axon into the spinal cord, even as far as the brainstem.

A very similar system allows **proprioceptors** to convey information concerning the position of the limbs and body, and the degree of tension in muscles. The axon of the nerve cell is located either in the muscle, tendon, or joint, and firing is instigated by pinching, as with touch receptors.

Nociceptors convey information about pain and include temperature, mechanical, and **polymodal** receptor types. Temperature nociceptors are activated only by extremely high or low temperatures. Mechanical nociceptors are activated by extremely strong pressure against the skin. Polymodal nociceptors are activated by high temperature, pressure, or chemicals released from damaged cells. Most nociceptors are free nerve endings unassociated with specialized membranes.

**Vision receptors.** Campaniform organs are insect proprioceptors that play a role in feedback control of body position, as well as normal walking motor programs that allow the coordinated movement of all six legs.

Vision receptors are called **photoreceptors** because the stimuli that activate them are photons of light. The two types of photoreceptors are called rods and cones. Rods only sense the intensity of light, while cones can sense both intensity and color. While cones function best in bright light, rods function better in dim light. Furthermore, rods are located diffusely over the **retina** at the back of the eye, but cones are located in the central line of vision in a region of the retina called the fovea. For this reason, dim objects in the darkness can be viewed better from peripheral vision than from direct focus. There are three kinds of cones in the vertebrate eye—one responsive to wavelengths of light corresponding to the color blue, one responsive to red wavelengths, and one responsive to green wavelengths. These three colors form the entire range of colors that humans can perceive.

Visual information is carried along the optic nerve and passes through several relay points before reaching the primary visual area of the cerebral cortex, located at the back of the brain. From there it splits into two routes. The dorsal pathway processes depth and motion while the ventral pathway processes color and form.

Insects have **compound eyes** organized in hundreds or thousands of practically identical units called ommatidia, each of which has its own lens system and its own tiny retina. This gives the eye a faceted appearance. The insect's retinula cell is a sensory neuron that contains photopigments similar to rods and cones. Overall, compound eyes resolve images about one sixtieth as well as vertebrate eyes. This is because the image is fractionated into so many ommatidia lenses and is difficult to reconstruct again in the brain. However, insects can have a greater range of vision than vertebrates, one that encompasses nearly the entire sphere of space around their bodies, due to the relatively large size of their eyes.

**compound eyes** multi-faceted eyes that are made up of thousands of simple eyes

**Hearing receptors.** Hearing receptors, or hair cells, are mechanoreceptors located within a bony spiral structure called the cochlea. Sounds are interpreted by the brain from patterns of air pressure caused by the vibration of objects. Sounds can also travel through water or solid objects. In mammals, the pressure in the air is transformed into mechanical pressure by three ear bones called the **malleus**, **incus**, and **stapes**, located in the middle ear.

Pressure waves that strike the tympanum, a thin membrane separating the middle from the outer ear, force it to push inward. The malleus is attached to the incus and the incus to the stapes, so that the mechanical activity of the tympanum is transferred to a coiled structure of the inner ear called the cochlea. Because this is a fluid-filled structure, the pushing and pulling of the stapes generates waves in the fluid. A semi-flexible membrane called the basilar membrane is located within the fluid, and also conducts the waves of pressure. The wave-like motion of the basilar membrane causes a series of hearing receptors grounded in the basilar membrane to be pushed up against another membrane just above it, the tectorial membrane.

**malleus** the outermost of the three inner ear bones

**incus** one of three small bones in the inner ear

**stapes** the innermost of the three bones found in the inner ear

Hair-like extensions, stereocilia, at the apex of the hair cell push and bend against a tectorial membrane, when the basilar membrane reaches the peak of its wave phase. This instigates a change in the electrochemical properties of the cell. The basilar membrane is formed so that only a particular region of hair cells is pushed up to the peak of the wave form for any one frequency, or tone, of sound. The frequency that a particular hair cell responds to is its receptive field. Hair cells are closely coupled to the auditory nerve, and transmit their auditory information to neurons from this nerve, which then travels up through the brain.

Insect hearing receptors are typically located on the legs, and sometimes on the body, of the insect. In crickets, the ear is located on the tibia of the foreleg. The eardrum of crickets is called a tympanum, and consists of an oval membrane on one side of the leg, and a smaller round tympanum on the other surface. Inside the tympanum is a large air sac formed by the leg trachea, which is the insect respiratory organ. Transduction mechanisms in the cricket ear are not well understood, but the vibration of the tympanum is known to stimulate the activity of auditory neurons behind the membrane.

**Balance receptors.** Vertebrate balance receptors are located in a specialized organ in the inner ear called the vestibular organ. This structure is located directly adjacent to the cochlea, and is composed of a triplet of semi-

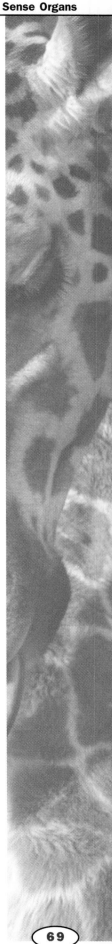

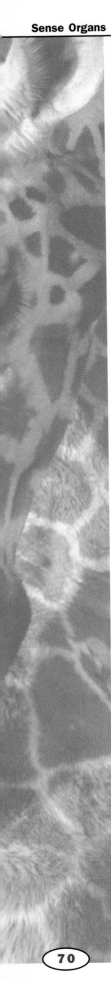

circular canals, each of which is oriented in a different plane—the X, Y, or Z axis. Movement of liquid in these tubes caused by rotation of the head or body are measured by vestibular hair cells. The stereocilia of these cells are embedded in a gelatinous material called the otolithic membrane.

Gravity and body movements cause the otolithic membrane to slide, which cause the stereocilia to bend in a particular direction. This leads to electrochemical changes in the hair cell, causes an action potential in the associated nerve ending. Information from the vestibular system allows eye and head movements to fix on a particular target, and to stabilize a moving image. It also allows organisms to balance—for example, when a cat walks atop a fence.

**olfactory** the sense of smell

**chemoreceptors** a receptor that responds to a specific type of chemical molecule

**Smell receptors.**   Smell receptors, or **olfactory** sensory neurons, are **chemoreceptors**, meaning that the binding of molecules causes these neurons to fire. Olfactory neurons extend a single dendrite to the surface of the skin in the nose, where it expands—along with dendrites from other neurons—to form a large knob. Thin hair-like projections extend from this knob into the thin layer of mucus within the nose. These projections contain a diverse array of receptors for odorants, so that all olfactory neurons are able to respond to a particular scent. The number that actually do respond is relative to the concentration of the scent molecules in the air.

**progeny** offspring

**Taste receptors.**   Taste-detecting, or gustatory, organs are also chemoreceptors and are located in functional groupings called taste buds on the tongue, palate, pharynx, epiglottis, and the upper third of the esophagus. Taste cells have a very short life span, which is why each unit contains a population of stem cells that continuously divides, producing **progeny** cells to replace the dying taste cells. The remainder of the cell types in the taste bud has hair-like projections called microvilli that extend into a pore at the top of the taste bud. When taste molecules bind or interact with the microvilli, the taste cell undergoes an electrochemical change that is conveyed to an associated neuron; however, taste cells are not neurons. Four basic taste sensations can be distinguished by humans: bitter, salty, sour, and sweet. SEE ALSO GROWTH AND DIFFERENTIATION OF THE NERVOUS SYSTEM; NERVOUS SYSTEM.

*Rebecca M. Steinberg*

**Bibliography**

Ehret, Gunter, R. Romand, et al. *The Central Auditory System.* New York: Oxford University Press, 1997.

Finger, Thomas E., D. Restrepo, Wayne L. Silver, et al. *The Neurobiology of Taste and Smell.* New York and London: Wiley-Interscience, 2000.

Hubel, David H. *Eye, Brain and Vision.* New York: Scientific American Library, 1995.

Kruger, Lawrence, ed. *Pain and Touch.* San Diego: Academic Press, 1996.

Singh, Naresh R., Nicholas James Strausfeld, et al. *Neurobiology of Sensory Systems.* New York: Plenum Press, 1989.

Whitlow, W. L., Arthur N. Popper, and Richard R. Fay. *Hearing by Whales and Dolphins.* New York: Springer, 2000.

# Serial Homology

As organisms evolve, their existing structures or body parts are frequently modified to suit their needs. For example, an invertebrate with a working limb design may end up changing it and incorporating it somewhere else in its **body plan**. The practice of modifying a specific structure more than once and using it somewhere else is known as **serial homology**.

As evolution is necessarily a stepwise process, certain complex structures, such as legs or wings, cannot spring into being instantly. They must slowly evolve over time, and each new and slightly different version must be more useful to its owner than the last. Replicating previously existing parts and building on them is a common strategy in organismal evolution. As such, serial homology is a widespread evolutionary tactic that can be observed in a large number of animals.

## Explaining Strange Mutations

Comparative biologists first had an inkling that copies of body parts were altered and used again when they started noticing odd mutants in their collections. At the end of the nineteenth century, comparative biologist William Bateson found some specimens in his collection that were odd-looking: he had **arthropods** with limbs in odd places, such as legs popping out of an animal's head. He also noticed certain ribs or vertebra swapped in other animals. However, these examples were rare and frequently unique. It wasn't until 1915 that Calvin Bridges, while breeding fruit flies (*Drosophila)*, came across a mutant that he could consistently breed where the rear flight appendage, or the haltere, resembled a wing.

This wing resemblance was no accident. The *Drosophila* wing and haltere are serially **homologous**. They were both modified from the same basic structure, and it should come as little surprise that interrupting the proper development pathway of one of them might cause it to resemble another. In terms of genetics, the two appendages are quite closely related; indeed, nearly identical. But at one location on the body, only a haltere will grow. At another, only a wing will grow. What mechanism makes this decision possible, the decision to grow a certain appendage on a certain part of the body? In the case of the wing and the haltere, the answer is **Hox genes**.

## Hox Genes

Hox genes are extremely common and evolutionarily very, very old. They are first **described** as belonging to a common ancestor of **bilateria** and **cnidaria** (in the neighborhood of 700 million years old). It is the Hox genes' job to locate different structures inside the organism's body plan. The particular gene in charge of making sure the haltere develops properly, and not into a wing or something else, is called *Ubx*.

What Bridges found was a partial mutation in *Ubx*. The halteres in his flies had wing bristles. *Ubx* controls a variety of other genes integral to haltere formation as well as genes important in the suppression of wing formation. *Ubx* discourages *spalt-related*, a gene that makes veins for the wings. It also stops genes that control wing epithelium formation, and other mechanisms and structures.

**body plan** the overall organization of an animals body

**serial homology** a rhythmic repetition

**arthropods** members of the phylum of invertebrates characterized by segmented bodies and jointed appendages such as antennae and legs

**homologous** similar but not identical

**Hox genes** also known as selector genes because their expression leads embryonic cells through specific morphologic developments

**described** a detailed description of a species that scientists can refer to identify that species from other similar species

**bilateria** animals with bilateral symmetry

**cnidaria** a phylum of aquatic invertebrates such as jellyfishes, corals, sea anemones, and hydras

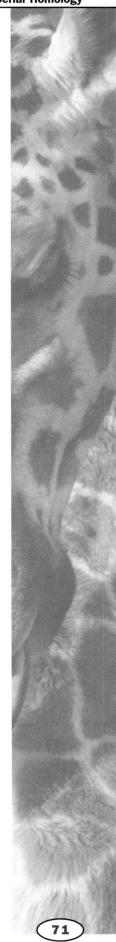

If there is a certain problem with the *Ubx* gene, a fly will grow halteres that have wing-like characteristics. While complete removal of a Hox gene generally results in death during early development, a certain triple mutation in *Ubx* can cause a second pair of fully-formed wings to develop where the halteres are supposed to be. Different mutations in Hox genes have produced flies with legs where antennae are supposed to be, and other odd body modifications.

## Hox Genes in Fruit Flies

In *Drosophila*, eight Hox genes, organized into two gene complexes, orient the body plan. By investigating where and when Hox genes were expressed, developmental biologists discovered that various genes were restricted to various body segments. Some of them overlap one another: *abd-A* and *abd-B* share a portion near the end of the abdomen, for example. The Hox genes in particular body segments affect the development of structures inside those segments. For example, the shape of the first pair of adult legs is influenced by the *Scr* gene, the second pair by the *Antp* gene, and the third pair by the *Ubx* gene.

While the Hox genes dictate the identity of a certain developing segment, they are not required for structures inside that segment to form. In *Drosophila*, the mouthparts and legs are serially homologous to the antennae. Thus, in the absence of these controlling Hox genes, the segments will still develop all their structures, but differently. Where the legs or mouthparts should be, antennae will develop instead. Such substitution of one part for another that is serially homologous is known as homeotic substitution. As William Bateson discovered with his collection, such substitution happens occasionally in nature to serially homologous structures.

So, as it is a useful evolutionary tactic to copy existing structures and modify them, the Hox genes evolved to keep tabs on what body segment is where so that the proper structural modifications can take place during development. While Hox genes are integral to differentiating the different segments of *Drosophila*, it is important to remember that they also delineate different areas of early developmental tissue (ectoderm, endoderm, and mesoderm) and specify location in a variety of fields for a staggering number of organisms.

There is a remarkable similarity in Hox gene sequences between *Drosophila* and vertebrates, such as mice, frogs, zebrafish, and humans. Alterations of Hox gene expression in *Drosophila* can prevent expression or growth of certain limbs or organs, and similar results have been found in birds, amphibians, and fish. This demonstrates that Hox genes function similarly over a wide variety of organisms, and that serial homology is a true evolutionary tactic when it comes to generating novel structures on one's body plan.

## Hox Genes and Body Designs

Different organisms use serial homology to different extents. Vertebrates have modified vertebrae and ribs in a large number of ways; the basic vertebra has been reiterated and altered into different backbone types: sacral, lumbar, **thoracic**, and cervical. However, perhaps one of the best examples

**thoracic** the chest area

of serial homology can be found in the body design of the common crayfish.

All appendages of crayfish, with the possible exception of the first antennae, are called biramous, which is to say they are derived from double-branched structures. The three components that make up these branches are known as the protopod (the base), the exopod (lateral; or on the side) and the endopod (in the middle). Crayfish have quite a large number of appendages in their body plans. The biramous structure plan is incorporated in crayfish head appendages, legs, swimmerets, mandibles, and many others.

Evolutionarily, it is easy to see how the earliest design might have been copied and modified several times, producing a highly specialized and versatile body plan. And as all the new structures are based on one older one, variations in Hox (or other developmental) genes will affect only a small number of structures, rather than all of them. Serial homology is a method for creating new, more specialized body plans, and it is observed in species throughout the living world. SEE ALSO MORPHOLOGY.

*Ian Quigley*

**Bibliography**

Carroll, Sean B., Jennifer K. Grenier, Scott D. Weatherbee. *From DNA to Diversity.* Malden, MA: Blackwell Science, 2001.

Davidson, Eric H. *Genomic Regulatory Systems.* San Diego, CA: Academic Press, 2001.

Hickman, Cleveland P. Jr., Larry S. Roberts, and Allan Larson. *Integrated Principles of Zoology*, 10th ed. Dubuque, IA: Wm. C. Brown Publishers, 1997.

# Service Animal Trainer

The Americans with Disabilities Act describes a service animal as any dog or other animal individually trained to provide assistance to a disabled human. Service animals perform some of the functions and responsibilities of a human without disabilities. For example, seeing eye dogs lead the blind through their environment, and hearing dogs alert their hearing-impaired owners to relevant sounds. Other examples include dogs that direct wheelchairs, fetch and carry items for mobility-impaired people, provide balance for unsteady people, monitor their owners for signs of a seizure, or provide therapeutic companionship.

Police dogs are service animals trained to recognize the scent of illegal substances such as drugs or gunpowder. They are also employed in locating missing persons, tracking down fugitives, and controlling jail riots. Search and rescue dogs are utilized on ski slopes, glacier parks, and mountains to seek out people who are injured or lost. They can be trained to rescue drowning people, pull a sled, deliver medications, or provide warmth for someone with hypothermia. Training teaches these dogs to be reliably calm and obedient in extreme situations.

Although not required, a background in animal behavior would benefit all animal trainers. Through a certification training program, or by apprenticing with a skilled trainer, anyone can acquire the qualifications necessary to train service dogs for impaired humans. The training program familiarizes students with concepts of classical conditioning, positive

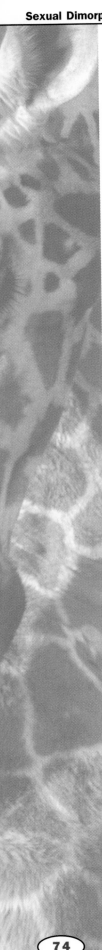

reinforcement, and operant conditioning, all of which involve reward, punishment, and emotional support for the animal.

Conversely, each state or country defines the requirements for becoming a police dog trainer. Most programs require a minimum amount of time as a police canine handler, completion of an instructor development program, written recommendation from other police canine trainers, additional coursework, and successful prior training of several service animals. Classes that offer certification the training of search and rescue dogs are open to anyone. The skill takes many years to master. In each of these cases, a strong relationship needs to form between the handler and the canine companion.

*Rebecca M. Steinberg*

### Bibliography

American Rescue Dog Association. *Search and Rescue Dogs Training Methods.* New York: Howell Book House; Toronto: Maxwell Macmillan Canada, 1991.

Duncan, Susan, and Malcolm Wells. *Joey Moses.* Seattle, WA: Storytellers Ink, 1997.

Robicheaux, Jack, and John A. R. Jons. *Basic Narcotic Detection Dog Training.* Houston, TX: J. A. R. Jons, 1996.

# Sexual Dimorphism

Sexual dimorphism is the presence of nongenital physical differences between males and females of the same species. These differences may be in coloration, body size, or physical structures, and can be quite striking. Male and female mallard ducks (*Anas platyrhynchus*) are so different in appearance that they were originally classified as different species. Mule deer bucks have antlers while females do not, male peacocks display an elaborate tail not found in females, and male elephant seals may weigh 2.5 tons (2.25 metric tons) more than their mates. All these differences are examples of sexual dimorphism.

The word "dimorphic" means "having two forms." Species in which males and females appear identical are called sexually monomorphic. Sexual dimorphism generally results from different reproductive roles and selective pressures for males and females. Scientists have proposed several hypotheses to explain the existence of sexual dimorphism, including niche **differentiation**, sexual selection through intrasexual competition, and sexual selection through female choice.

## Niche differentiation

The niche differentiation hypothesis states that the differences between males and females of a species allow them to exploit different food resources and thus reduce competition for food. For example, male raptors (hawks, falcons, eagles, and owls) are considerably smaller than females. This difference is especially pronounced in raptors that prey on birds, such as the Cooper's hawk (*Accipiter cooperii*). The small male Cooper's hawks are better than their larger mates at catching small, fast birds. The male catches these often-abundant small birds to feed to his mate as she incubates the eggs and later while the chicks are young. Once the young hawks are bigger, the female hunts for larger birds in the area to feed the growing chicks.

**differentiation** differences in structure and function of cells in multicellular organisms as the cells become specialized

## Intrasexual Competition

In 1871 Charles Darwin suggested that larger, more aggressive males would be more successful in competing for females. Intrasexual competition results in males that are larger, stronger, or equipped with different physical traits than females because they must compete with other males to win access to mates. American elk (*Cervus elaphus*) bulls have large antlers and use them aggressively to fight for harems of females. A harem is a group of adult females defended by one male from other males in the area. This dominant male mates with most if not all females in the group. Bulls that have large antlers and outweigh their opponents are able to dominate their opponents and pass on their genes for size to their offspring. Elk cows are not involved in such battles and have no need for antlers or such massive bulk.

Finally, the female choice hypothesis states that females of dimorphic species prefer to mate with males that are colorful, large, or ornamented. If this is the case, the flashier, larger, or more-ornamented males are more likely to sire offspring and pass on their genes than their dull, small, plain competitors. Why females might prefer to mate with elaborate males is a matter of debate. It may be that the traits advertise the males' ability to find food or avoid predators.

## Sexual Selection

By mating with obviously successful males, females increase the chances that their offspring will inherit their father's traits and also be successful. It may also be that some sexually selected traits, such as the peacock's unwieldy tail, advertise the male's ability to escape predators in spite of a severe handicap. Longer tails make it harder for the males to run or fly away, so males that have particularly long tails and yet manage to escape predators must be particularly strong and fast. Competition between males, and female choice among males are both forms of sexual selection. This means that they cause differences in mating success of individuals and therefore result in adaptations for obtaining mates.

In species that are sexually dimorphic, the male is usually the bigger, more brightly colored, or more elaborately ornamented sex. However, there are several cases where this pattern is reversed, such as in raptors and toads. The female is often the larger sex in polyandrous birds such as phalaropes, where brightly colored females compete to attract male mates. This unexpected situation, sometimes called reverse sexual dimorphism, led the American ornithologist and painter John James Audubon (1785–1851) to incorrectly label the male and female in all his phalarope paintings. The plumage patterns of phalaropes make sense in light of sexual selection, but also highlights the importance of selection for dull plumage. Male phalaropes incubate the eggs and take care of the young with no help from the female. Their cryptic (meaning blending into the surroundings) brown plumage makes them difficult to see on the nest and helps them avoid predators. In general, the sex that cares for the young is under strong selection for dull coloration.

The hypotheses discussed here are not mutually exclusive, and different species may be more or less affected by various selective pressures. Why

A male mandrill rests on his paws at the Ft. Worth, Texas, zoo. Male mandrills exhibit coloring consistent with the theory of sexual dimorphism.

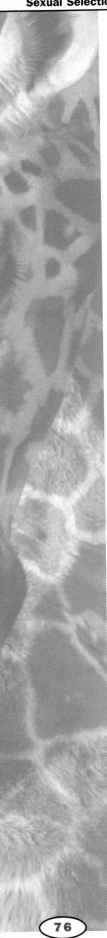

some species are sexually dimorphic and others are not is an active area of research in evolutionary biology.

*Emily DuVal*

**Bibliography**

Darwin, Charles. *The Descent of Man, and Selection in Relation to Sex.* London: John Murray, Albermarle Street, 1871.

Ehrlich, Paul R., David S. Dobkin, and Darryl Wheye, eds. *The Birder's Handbook: A Field Guide to the Natural History of North American Birds.* New York: Simon and Shuster, Inc., 1988.

Owens, Ian P. F., and Ian R. Hartley. "Sexual Dimorphism in Birds: Why Are There So Many Different Forms of Dimorphism?" *Proceedings of the Royal Society of London*, series B. 265 (1998):397–407.

Pough, F. Harvey, Christine M. Janis, and John B. Heiser. *Vertebrate Life*, 5th ed. Upper Saddle River, NJ: Prentice Hall, 1999.

Zahavi, Amotz, and Avishag Zahavi. *The Handicap Principle: A Missing Piece of Darwin's Puzzle.* New York: Oxford University Press, 1997.

# Sexual Selection

Males and females of many species exhibit significant differences in addition to the difference in reproductive organs. The distinction of gender through secondary sex characteristics is known as sexual dimorphism. This is most often expressed as a difference in size, with the males usually larger, but also involves differences such as plumage in male birds, manes on male lions, and antlers on male deer. In most cases, the male is the showier sex of the species.

## What It Is and How It Works

Sexual selection is the evolutionary process that arises from competition among members of one sex (the competitive sex) for access to members of the other sex (the limiting sex). According to the theories of Charles Darwin, sexual selection should be distinguished from the process of natural selection because the traits that evolve via sexual selection often appear to have a negative effect on the survival rate of their bearers. There are two basic forms of competition for mates that affect the type of traits that evolve. One is intrasexual selection, which includes overt competition among the members of the competitive sex to gain control or monopoly over the limiting sex. This leads to selection of traits such as weapons, large body size, aggression, strength, and endurance. Intersexual selection is when the limiting sex can exercise a choice of mates, leading to elaboration of structures, displays, **vocalizations**, and odors in the competitive sex.

**vocalizations** sounds used for communications

Sending and receiving signals is a significant step toward mating as the type and intensity of the signal will determine whether or not a mate is obtained. One process of signal evolution that is involved in the search for a mate is called sensory exploitation. According to this model, signal receivers often have inherent, or built-in, preferences that can be exploited by a manipulative signaler to create new signals. For instance, suppose female birds searched preferentially for red seeds while foraging. Because the only time they encounter red is in seeds, it would be advantageous to evolve a general preference for red objects. A mutant male that adds red to its plumage may

This male prairie chicken displays his colors, indicating he is ready to breed.

then be able to exploit this preference for the color red as long as it can be expressed in the context of mate choice. A new signal can then evolve that had no historical link to mate choice but only to an irrelevant context such as foraging.

It is important to note that there is no new information conveyed to the females by the red males. In fact, there may be predatory risks if red males are easier to spot and thus more likely to be killed. Costs to females may be an adaptation to avoid exploitation such as better discriminating abilities or the decoupling, that is, the separation of behavioral strategies, for foraging and decisions about mating. However, females may be able to find males of the same species, termed conspecific males, more easily if the males bear a red patch. Furthermore, red males may provide additional information not provided by normal males. If the intensity of redness is a good indicator of the males' health, then a preference for red males over nonred ones could be a way for females better to identify a good mate. Whether the signal is costly or beneficial, sensory exploitation by itself is highly unstable. Instead, it is followed by **coevolution** between sender and receiver.

## Male Attraction Signals

Females are more often the choosy sex and males the competitive, advertising sex (though in some animals the roles are reversed). Two significant models for the evolution of mate-attraction signals are the *runaway selection model* and the *good genes model*. In both of these there is a simultaneous evolution of a female preference for a particular male trait and evolution of that male trait.

**Runaway selection model.** This model best defines polygynous species (one in which a number of mates are taken). Initially, there must be some genetic variability associated with variability in physical characteristics of a male trait such as the brightness of a color spot. Males with the pre-

**coevolution** a situation in which two or more species evolve in response to each other

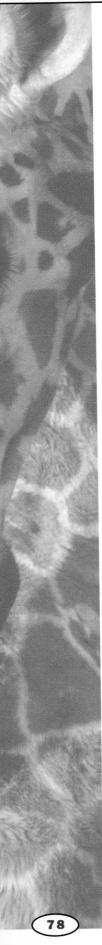

ferred trait will obtain more mates, and because the females will be those who tend to prefer that trait, the genes for the female preference will become linked with the male trait in their offspring. Females with the preference also benefit because their sons will have that trait and so will win more mates, thereby spreading the genes of the female. Under the right conditions, a runaway evolutionary process can occur with the male trait becoming larger and the female preference for it becoming stronger. This process stops when the male trait becomes so large that it imposes a cost on the male that outweighs the benefits. The traits that evolve through this model are termed arbitrary traits, because they can take any type of conspicuous form and not provide any information to the female about the male's fitness.

Although this process requires a certain set of circumstances to get under way, there are several different situations that could bring about the initial trait and the preference for it. One possibility is sensory exploitation, as discussed above. Another factor could be that the male trait is initially favored by natural selection and a female preference for this trait subsequently evolves. A third possibility is that the females evolve a preference for the trait because it helps distinguish conspecific males from those of a closely related species, thereby preventing genetic mixing between different species, a process termed interspecific hybridization.

**Good genes model.**    This model actually proposes that costly and conspicuous male traits become the trait of choice by females because they indicate some aspect of male quality. Females will benefit from mating with these males because the offspring will have higher survivorship or viability. The male traits are called indicator traits, and the costs to males are deemed necessary costs when in pursuit of a mate. Coevolution of the male indicator trait, intrinsic male viability, and female preference for the trait is the basis for the genetic model of this process. One example of how this may work is the *classic handicap model*, in which males acquire a trait such as a heavy set of antlers, which obviously imposes a survivorship cost. Low-quality males cannot support the cost of this trait, and so it is seen as a handicap, whereas high-quality males survive and their genes dominate. However, the handicap is balanced out by the higher net fitness of males without this trait and its cost. Another theory is the *condition-dependent indicator model*, in which males vary the expression of the preferred trait, such as the speed at which the call is performed (termed call rate) so as to optimize their mating success and survival. High-quality males can afford to expend more energy on expression than can low-quality males and so trait magnitude is a good indicator of male quality. Last, in the revealing indicator model all males attempt to develop the trait to the same magnitude and pay the same cost, but the condition of the trait is lower in low-quality males. For instance, call frequency (the number of times a call is repeated) or feather condition are good examples of such traits.

## Signalling for a Mate

Many aspects are taken into account when an animal signals for and chooses a mate. Anatomical traits play a role in the attraction of a potential mate. Non-weapon body structures such as color patches, elongated tails, and fins and feather plumes are termed ornaments. Many ornaments appear to be good indicators of the quality of the male. They often come into play when the male performs visual displays to attract the attention of a female. Vigorous displays are indicative of a male's high energy, and thus that male is likely to win out over the less vigorous male.

**Auditory signals.** In birds, insects, mammals, and fish, these are also obvious targets of female choice. In all of these groups, the females tend to prefer males with greater calling rates, sound intensity, and call duration. These features actually increase the stimulation value of the signal, so it is possible that the preference arose from the sensory bias of the female receivers and that the traits themselves are just arbitrary. These call characteristics are all energetically expensive, and it seems that females prefer the most costly calling. For example, female grey tree frogs prefer calls that have a long duration and are repeated at a slow rate over a call that is brief and at a high rate though it has the same acoustic "on" time, meaning the calls are heard for the same amount of time. Studies have shown that the long duration calls require a lot more energy than the shorter calls. Such findings suggest that display vigor is an accurate indicator of male quality. It has also been found that costly call characteristics are correlated with age, size, dominance, or parasite load. In addition, they provide females with good gene benefits such as faster growing and more viable offspring. Finally, in some species, call rate is a good index of benefits the females can expect to receive, such as large sperm or territory rich in resources.

**Vocal signals.** Another type of call is the **copulation** call. These are usually vocal signals and may be given by the male, female, or both sexes. In most cases it is difficult to determine the function of the signal and the intended receiver. However, copulation signals are unlikely to be incidental as they are highly structured and individually distinctive. In monogamous species, female copulation signals may be for synchronization of orgasm with the male, as there is usually no other possible intended receiver in the vicinity at that time. In socially mating species, the intended receiver could be external to the copulating pair. The female signals could be intended for other females and to increase dominance status in the group. The signals could also serve as a recruiting call for other male mates and incite competition among them so that the female can choose the dominant male as father of her offspring.

**copulation** the act of sexual reproduction

Male copulation signals could serve to transmit information about the male's mating success to other females. It has been found that males who make these calls are more likely to get attacked by other males, but they are also more likely to obtain more matings compared to males that 'do not call. Alternatively, the intended recipient of male copulatory signals could be other males. For example, postcopulatory male rats repeatedly **emit** vocalizations that are similar to the ultrasonic whistles signaling an alarm situation or a defensive threat. The male rats often appear lethargic and inactive at this time but will aggressively attack any other male that tries to approach the mated female. The signal then seems to be indicating mate-guarding intentions.

**emit** to throw or give off

**The nuptial gift.** This is a slightly different tactic in the search for a mate. Males of some insect and bird species will offer prey items as nuptial gifts during courtship. The females may base their decision on the size,

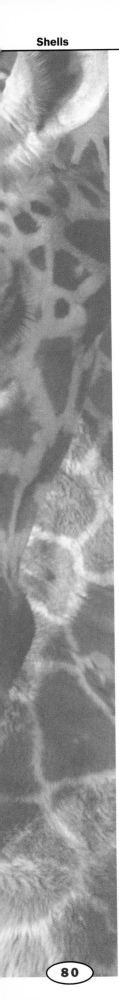

or quality of these items, as well as the rate at which these items are provided. Females that are egg-producing would clearly benefit from receiving food as a nuptial gift. If provisioning rate is correlated with offspring feeding rate in paternal care species, the behavior is an obvious indicator with direct benefits. Another consideration is that if the cost to the male increases with gift magnitude, then it could be an indicator of heritable fitness. However, there is a risk of "false advertisement." In the marsh hawk, provision of the nuptial gifts is a good indicator of the male's nesting provisional ability. But the males sometimes use this signal deceptively in order to attract females into polygynous matings, a disadvantageous position for the female hawk.

In some species there is a sex-role reversal in terms of competing for mates. The sex-role reversal results from evolution of male parental investment, although not all paternal care species display complete reversal of courtship roles. A determining factor is the extent to which the male cares for the offspring. If the males can care for the offspring of several females simultaneously, the sex ratio is still skewed in favor of males. In those circumstances, males are still the competitive gender and will perform aggressive or persuasive courtship behaviors. However, if the males can care for the offspring of only one female, then males become a limiting resource for females. The females will then compete for males and develop aggressive behaviors, ornaments, and mate-attraction displays most often associated with the male gender. SEE ALSO REPRODUCTION, ASEXUAL AND SEXUAL; SEXUAL DIMORPHISM.

*Danielle Schnur*

**Bibliography**

Andersson, M. *Sexual Selection*. Princeton, NJ: Princeton University Press, 1994.

Bradbury, Jack W., and Sandra L. Vehrencamp. *Principles of Animal Communication*. Sunderland, MA: Sinauer, 1998.

Campbell, Neil A. *Biology*, 3rd ed. Berkeley, CA: Benjamin/Cummings Publishing Company, 1993.

Parker, G. A. *Mate Choice*. Cambridge, U.K.: Cambridge University Press, 1983.

## Sharks  *See Chondricthyes.*

# Shells

Every year, thousands of shells wash up on beaches around the world. Have you ever found a shell and wondered where it came from or what its function once was?

These shells come from a large phylum of animals known as mollusks which includes members such as snails, clams, oysters, scallops, squid, and octopuses. One of the prominent characteristics of most mollusks is a hard exterior shell, although some mollusks, including the squid and the sea hare, produce internal shells.

The bodies of mollusks are very soft. Their shells protect them from predators. The body of the mollusk is formed of a combined structure called a "head-foot" that is used for locomotion. If threatened, shelled mollusks can retreat the "head-foot" quickly inside their shells.

A thin, fleshy fold of tissue called the **mantle** covers the internal organs of most mollusks. The mantle, which is made up of specialized cells, secretes the substance that creates the shell. The shell is formed almost entirely of calcium carbonate crystals, which is the chemical base of rocks like limestone. The shell starts as a semi-liquid substance that hardens very rapidly as soon as it is exposed to water or air.

The shell begins to form as tiny crystals of calcium carbonate are exuded onto a **matrix** of **conchiolin**, a brownish, horn-like material composed of proteins and **polysaccharides** produced by the animal. This surface serves as a microscopic latticework for new crystals. Additional calcium carbonate crystals are laid down in thin layers as either **calcite** or **aragonite**.

Shells vary in structure and hardness depending upon their crystallization type. Six-sided calcium carbonate crystals may be prism-like, becoming calcite, or they may be very flat and thin, becoming aragonite.

Both calcite and aragonite are forms of calcium carbonate. Aragonite is heavier than calcite, and it has a mother-of-pearl appearance. The shells of freshwater clams and land snails are typically made of aragonite. In saltwater genera such as Nerita, layers of calcite and aragonite may alternate during shell formation. Whether shells are made of calcite or aragonite, as additional layers are added, the shells grow in thickness.

Many mollusks also have special glands called chromogenic glands located on the margin of the mantle. These glands secrete colored pigments that stain the calcium carbonate, which would otherwise create a white shell. When the glands are in a continuous series and are steadily active, the developing shell will be a solid color. If the glands secrete continuously but are some distance from each other, spiral lines will appear on the shell. The color will show as dotted lines, if the glands work intermittently. In some cases, groups of adjacent glands secrete different pigments, creating multicolored shell designs of infinite variety.

Shelled mollusks are linked directly to their shells; they ordinarily cannot live separately from the shell. When a mollusk dies and the shell is no longer attached to a living creature, other aquatic animals (i.e. crustaceans such as crabs) sometimes use the shell as temporary shelter, in the water or on the shore. Throughout human history, shells have been valued as decorative objects, tools, and mediums of economic exchange. They continue to be popular collectibles around the world. SEE ALSO MORPHOLOGY.

*Stephanie A. Lanoue*

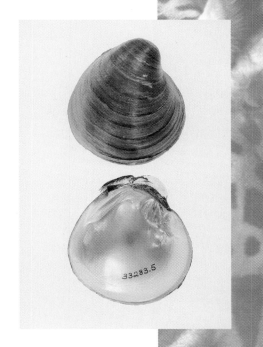

A ring pink mussel.

**mantle** the tissue in mollusks that drapes over the internal organs and may secrete the shell

**matrix** the nonliving component of connective tissue

**conchiolin** a protein that is the organic basis of mollusk shells

**polysaccharides** carbohydrates that break down into two or more single sugars

**calcite** a mineral form of calcium carbonate

**aragonite** a mineral form of calcium carbonate

**Bibliography**

Saul, Mary. *Shells.* New York: Doubleday and Company, Inc., 1974.

Solem, Alan. *The Shell Makers.* New York: John Wiley and Sons, 1974.

Tucker, Robert. *Kingdom of the Shell.* New York: Crown Publisher, Inc., 1982.

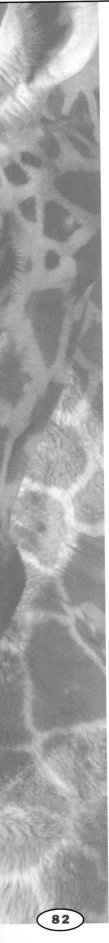

# *Silent Spring*

In 1955 Rachel Carson was at the peak of her profession as a popular writer of science books about the sea. *Under the Sea Wind* (published in 1941), *The Sea Around Us* (1951), and *The Edge of the Sea* (1955) were all best-sellers, and they catapulted her into celebrity as one of the best loved and most sought after American authors. By 1956, she was planning a book in which she intended to explore the human race's relationship with nature. Fearing that human beings were severing their connection to the web of life, she began the painstaking research that would form her next work, *Silent Spring*, a book that would change her life and the world.

Carson's deepest held beliefs were in the delicate interconnectedness of nature and the sanctity of life. Her work as a biologist for the U.S. Fish and Wildlife Service during World War II (1939–1945) had involved studying documents concerning the horrendous chemical and human-made devastations that were occurring. After the war, many of the chemicals developed for the military were unleashed on neighborhoods and farms in a war against nature. By 1960, there were some 200 untested chemicals used in pesticide formulas. That same year, 638 million pounds of poisons were broadcast in the United States alone. The chemical pesticide business was a $250 million industry enthusiastically supported by the U.S. Department of Agriculture (USDA) and other agencies. Government researchers had documented the dangers of uncontrolled pesticide use, but their warnings were ignored or destroyed, with many of the scientists encouraged to find other jobs. A chance note from an ornithologist friend concerning a die-off of baby birds after a DDT spraying of a nearby marsh spurred Carson to act. As she began reading hundreds of scientific papers and contacting biologists, chemists, agriculture experts, and doctors around the world, her alarm and her determination grew. "The more I learned about the use of pesticides the more appalled I became. I realized that here was the material for a book. What I discovered was that everything which meant the most to me as a naturalist was being threatened and that nothing I could do would be more important" (Jezer 1988, p. 79).

After four years of research, *Silent Spring* began appearing in a serialized condensed version in *The New Yorker* on June 16, 1962. It included an appendix of more than fifty pages of scientific references. Response was immediate and overwhelming. Praise and concern in the form of thousands of letters and telegrams poured into the magazine from citizens, scientists, and even the new U.S. president, John F. Kennedy. The response from the American Medical Association, the USDA, and the chemical companies was even more vocal, however. They targeted a quarter of a million dollars for a brutally negative publicity campaign, impugning Carson's science and her morals. One member of a government pest control board scoffed that Carson had no business worrying about genetics as she was a "spinster." The Velsicol Chemical Corporation sent her publisher a threatening letter insisting that Carson was part of a communist conspiracy to undermine the economy of Western nations. Houghton Mifflin was undeterred and the book was published on schedule, on September 27, 1962. As the book soared on the best-seller list, the attacks intensified in print and on television. Her opponents must have realized—as was indeed the case—that she was questioning not only the indiscriminate use of poisons but also the basic irresponsibility of an industrialized, technological society toward the natural world. She refused to accept the premise that damage to nature was the inevitable cost of progress.

President Kennedy initiated a Science Advisory Committee to study the dangers and benefits of pesticides. After eight months of study, their report concluded that "Until the publication of *Silent Spring* by Rachel Carson, people were generally unaware of the toxicity of pesticides." A U.S. Senate committee was formed to study environmental hazards. Secretary of the Interior Stewart Udall acknowledged, "She made us realize that we had allowed our fascination with chemicals to override our wisdom in their use." Most importantly she touched a chord in the population of the United States and the dozens of countries worldwide where her book was translated. Grassroots conservation and environmental organizations sprang up demanding political action. By the end of 1962 more than forty bills regulating pesticides had been introduced in legislatures across the United States. By 1964, the U.S. Congress had amended federal laws to shift the burden of demonstrating the safety of new chemicals to the manufacturers, requiring the proof of safety before the chemicals could be released. As her ideas gained momentum, Carson was showered with honors and awards, including the Audubon Medal and honors from the American Academy of Arts and Letters, the National Wildlife Federation, the Animal Welfare Institute, and the American Geographical Society. Carson continued to warn that "modern science has given human beings the capacity to destroy in a few years life forms that have taken eons to evolve. Humans are challenged to use this new power intelligently and cautiously. Conservation is a cause that has no end" (Jezer 1988, p. 99).

On April 14, 1964, Carson died of breast cancer. Before she wrote *Silent Spring* few people were aware of the ecological principle that all of life is interrelated. Because of her courage, determination and eloquence, these ideas have become widespread. Millions of human beings have begun to take responsibility for humanity's place in the natural world. They agree with Carson that "man is a part of nature and his war against nature is a war against himself. The human race now faces the challenge of proving our maturity and our mastery, not of nature, but of ourselves" (Jezer 1988, p. 105). SEE ALSO CARSON, RACHEL; DDT; ENVIRONMENTAL DEGRADATION; HABITAT; HABITAT LOSS; PESTICIDE.

*Nancy Weaver*

**Bibliography**

Carson, Rachel. *Silent Spring*. Boston: Houghton Mifflin, 1962.

Harlan, Judith. *Sounding the Alarm*. Minneapolis: Dillon Press, 1989.

Jezer, Marty. *Rachel Carson*. New York: Chelsea House, 1988.

# Silurian

The Silurian period of the Paleozoic era, 440–410 million years ago, follows the Ordovician period of the Paleozoic, "the age of ancient life." The Silurian was named by the R. I. Murchison in 1835 in honor of the Silure

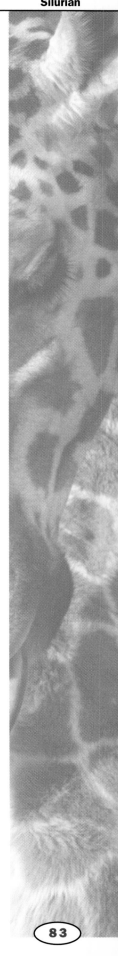

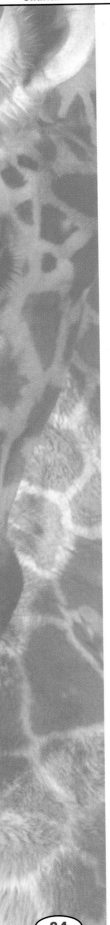

The silurian period and surrounding time periods.

| Era | Period | Epoch | Million Years Before Present |
|---|---|---|---|
| Paleozoic | Permian | | 286 |
| | Pennsylvanian | | 320 |
| | Missipian | | 360 |
| | Devonian | | 408 |
| | Silurian | | 438 |
| | Ordovician | | 505 |
| | Cambrian | | 570 |

tribe of Celts who had inhabited the Welsh borderlands where he first studied these rocks.

A long, warm, stable period followed the ice age and mass extinctions of the Ordovician, when over half of all previous life forms became extinct. Silurian fauna built upon the evolutionary patterns that had preceded it. No major new groups of invertebrates appeared, although a great radiation in number and form of existing invertebrates occurred. It has been said that the evolution of life cannot be separated from the evolution of the planet. In the Silurian, geologic trends greatly influenced animal development. As the two great supercontinents, Laurasia in the Northern Hemisphere and Gondwanaland in the Southern Hemisphere, once again drifted toward one another on their tectonic plates, mountains were heaved up to form distinct ecosystems where species could evolve uninfluenced by one another. And as the glaciers began to melt, warm, shallow seas flooded much of Laurasia, providing ideal conditions for a variety of **benthic** (bottom-dwelling) species. These included a rich variety of sea lilies, lampshells, trilobites, graptolites, and mollusks. The crinoidal sea lilies and graptolites are particularly interesting. They are both echinoderms: small, soft-bodied wormlike creatures that lack a normal head but have a well-developed nervous system located in a rudimentary **notochord** ("backchord"), the precursor of a backbone. The presence of a notochord makes these once-abundant sea-floor scavengers the ancestors of the chordates, animals with backbones.

A recurrent theme occurs in the origin of new marine species. They first tend to appear in very shallow waters along the shore, then disperse into deeper habitats. The shoreline is a harsh area of constant tides, storms with silt flows, and temperature fluctuations. These conditions favor species that are resilient and adaptive. Gradually the offspring expand into deeper water. The new forms of all the existing invertebrates followed this pattern: the **brachiopods**, sponges, bryozoans, **arthropods**, and echinoderms, as well as the vertebrate fishes. In the deeper waters, mobile predators appeared in unprecedented sizes. The free-swimming nautiloids, which grew up to 3 meters (10 feet) long and the eurypterids, sea-scorpion arthropods at 2-meters (6-feet) long, fed on the vast numbers of early jawed fishes that now appeared. The great reefs destroyed by the ice age were rebuilt, coral by individual coral.

The most noteworthy event of the Silurian (from the human point of view) took place on land. The first minuscule plants began to creep across the previously barren land masses, followed by tiny scorpions and millipedes. The whiskery, or pleated, tracks of arthropods appear in the Silurian rocks of western Australia, and for a brief while these arthropods dominated Earth. The formerly rare agnathans (jawless fishes), became plentiful and began to

**benthic** living at the bottom of a water environment

**notochord** a rod of cartilage that runs down the back of Chordates

**brachiopods** a phylum of marine bivalve mollusks

**arthropods** members of the phylum of invertebrates characterized by segmented bodies and jointed appendages such as antennae and legs

explore up the **brackish estuaries** and into the freshwater rivers and up-stream pools where they flourished. SEE ALSO GEOLOGICAL TIME SCALE.

*Nancy Weaver*

**Bibliography**

Asimov, Isaac. *Life and Time.* Garden City, NY: Doubleday & Company, 1978.

Fortey, Richard. *Fossils: The Key to the Past.* Cambridge, MA: Harvard University Press, 1991.

———. *Life: A Natural History of the First Four Billion Years of Life on Earth.* New York: Viking Press, 1998.

Friday, Adrian and David S. Ingram, eds. *The Cambridge Encyclopedia of Life Sciences.* London: Cambridge University, 1985.

Gould, Stephen Jay, ed. *The Book of Life.* New York: W. W. Norton & Company, 1993.

Lambert, David. *The Field Guide to Prehistoric Life.* New York: Facts on File, 1985.

McLoughlan, John C. *Synapsida: A New Look Into the Origin of Mammals.* New York: Viking Press, 1980.

Steele, Rodney and Anthony Harvey, eds. *The Encyclopedia of Prehistoric Life.* New York: McGraw Hill, 1979.

Wade, Nicholas, ed. *The Science Times Book of Fossils and Evolution.* New York: The Lyons Press, 1998.

# Simpson, George Gaylord

### *American Paleontologist and Biologist*
### *1902–1984*

George Gaylord Simpson was born to middle-class parents in Chicago, Illinois, on June 16, 1902. The family soon moved to Colorado, where he became fascinated by the dramatic geology and vertebrate fossils of the West. Graduate work at Yale University culminated in a Ph.D. in Geology in 1926 with a thesis on Mesozoic mammals. Simpson married psychologist Anne Roe and later collaborated with her on several books and conferences about behavior and evolution.

In 1927 Simpson began his thirty-two-year association with the American Museum of Natural History, of which he became curator in 1942. During those years he led expeditions to Mongolia, Patagonia, and Montana and taught at Columbia University. His work in organizing all the known fossil vertebrates of the Mesozoic, Paleocene, and Eocene was summarized in a series of textbooks including: *Tempo and Mode of Evolution* (1944), *The Meaning of Evolution* (1949), and *The Major Features of Evolution* (1952). His most significant achievement may be the application of population genetics to the analysis of the migration of extinct mammals between continents.

From 1959 to 1970 Simpson was professor of vertebrate paleontology at the Museum of Comparative Zoology at Harvard University. In 1967, at the age of sixty-five, he moved to Tucson to become professor of geosciences at the University of Arizona.

Simpson's lifelong enthusiasm for and contributions to his chosen field were recognized by numerous honorary degrees and medals worldwide. He was elected to the National Academy of Sciences in 1941 and the National Academy of Arts and Sciences in 1948. He was cofounder and first presi-

**brackish** describes a mix of salt water and fresh water

**estuaries** areas of brackish water where a river meets the ocean

George Gaylord Simpson's work has helped generations of biology students to better understand the types of animals that lived through the Mesozoic, Paleocene, and Eocene time periods.

The skeleton of a turkey.

**bone tissue** dense, hardened cells that makes up bones

**hydrostatic skeletons** pressurized, fluid-filled skeletons

**cartilage** a flexible connective tissue

**connective tissues** cells that make up bones, blood, ligaments, and tendon

**lancelets** a type of primitive vertebrate

**cartilaginous** made of cartilage

**osteoblasts** potential bone-forming cells found in cartilage

**ossification** the deposition of calcium salts to form hardened tissue such as bone

**axial skeleton** the part of the skeleton that makes up the head and trunk

**appendicular skeleton** the part of the skeleton with the arms, legs, and hips

dent of both the Society of Vertebrate Paleontologists and the Society for the Study of Evolution. George Gaylord Simpson died October 6, 1984.

*Nancy Weaver*

### Bibliography

Greene, Jay, ed. *Modern Scientists and Engineers*, vol. 3. New York: McGraw-Hill, 1980.

Porter, Ray, ed. *The Biographical Dictionary of Scientists*. New York: Oxford University Press, 1994.

Simpson, George Gaylord. *Attending Marvels: A Patagonian Journal*. New York: Time-Life Books, 1965.

# Skeletons

Skeletons provide the framework for the bodies of most multicellular animals. They lend structural support to soft tissues and give muscles something to attach to and pull against. Without skeletons, most animal bodies would resemble a limp bag of gelatin.

Skeletons come in a number of forms, each suited for a particular set of lifestyles and environments. Skeletons can be rigid, semirigid, or soft. They can also be external or internal. Vertebrates have internal skeletons, called bony skeletons, which consist mainly of calcified **bone tissue**. Most invertebrates, such as insects, spiders, and crustaceans, have outer skeletons called exoskeletons. Some aquatic animals, such as octopuses, sea anemones, and tunicates, and a number of small, land-dwelling invertebrates such as earthworms and velvetworms, have soft supporting structures called **hydrostatic skeletons**.

## Bony Skeleton

*Vertebrata* (vertebrates) is an animal group that includes fishes, amphibians, reptiles, and mammals. The vertebrate skeleton is an internal collection of relatively rigid structures joined by more flexible regions. The hard components of the skeleton are made up of bone, **cartilage**, or a combination of these two **connective tissues**.

Vertebrates are closely related to a number of less-familiar aquatic organisms, such as tunicates, sea squirts, and **lancelets** (Amphioxus). These animals have a skeleton composed entirely of a **cartilaginous** rod called a notochord. The notochord is somewhat flexible and runs along the back of the animal.

In all vertebrates, the framework first laid down during development is cartilaginous. As development proceeds, most of the cartilage is replaced by calcified bone through the action of bone precursor cells called **osteoblasts**. This process is called **ossification**. During ossification, some bones fuse together, reducing the total number of bony elements. At birth, human infants have over 300 bones. As adults, they have 206.

The bony skeleton of vertebrates consists of an **axial skeleton** and an **appendicular skeleton**. The axial skeleton is made up of the skull, spinal column, and ribs. This skeleton provides the general framework from which the appendicular skeleton hangs. The appendicular skeleton consists of the pelvic girdle, pectoral girdle, and the appendages (arms and legs).

Bony skeletons have a number of advantages over other types of skeletons. Because bony skeletons are living tissue, they can grow along with the rest of the body as an individual ages. As a result, animals with bony skeletons do not replace their skeletons as they grow older. Bone itself is a dynamic tissue that adjusts to the demands imposed by its environment and by its owner. Bone not only repairs itself when broken, but thickens in response to external stresses.

Bony skeletons are denser and stronger than exoskeletons and hydrostatic skeletons, and are able to support animals of a large size. By assuming a more upright posture, large animals can support a tremendous amount of weight on their skeletons. Internal skeletons are also less **cumbersome** in large animals than external skeletons would be. As an animal increases in size, its surface expands to an area that would be too large to be reasonably accommodated by an exoskeleton. All large, land-dwelling animals have bony skeletons.

**cumbersome** awkward

Bony skeletons can respond to increasing weight-bearing demands by adjusting bone density and by distributing the weight through changes in posture. However, the bones of some animals have actually become lighter to accomodate other functions. Bird bones, which are hollow structures, constitute a mere 4 percent of the animal's body weight, compared to 6 percent in the mammals.

A major disadvantage to the bony skeleton, relative to an exoskeleton, stems from its internal location. Although certain elements of the bony skeleton, such as the skull and rib cage, provide protection to the soft organs they encase, the skeleton offers no protection to the other soft tissues of the body. External protection is therefore left to other structures, such as the skin and its associated hair, fur, and nails.

## Exoskeleton

Exoskeletons are found in most invertebrates and assume a variety of forms. Some exoskeletons are made of calcium or silica, as seen in protozoa called foraminiferans. Exoskeletons can also be elastic, such as those worn by sponges, or hard and stony, like those secreted by coral. In contrast, mollusks (clams and snails) house themselves in hard shells comprised mainly of calcium carbonate.

When compared to bony skeletons, exoskeletons have two advantages. First, they provide a hard, protective layer against the environment and potential predators. And second, they protect their wearer against drying out, which is a great threat to land-dwelling species. It is important to avoid **dessication** because water molecules play an important role in many of life's critical physiological processes, including those related to digestion and circulation.

**dessication** drying out

**chitin** a complex carbohydrate found in the exoskeleton of some animals

Insect, spider, and shellfish (lobster and shrimp) exoskeletons contain a compound called **chitin**, a white horny substance. These **arthropods** have segmented exoskeletons that bend only at the joints. The exoskeleton covers the entire surface of an arthropod's body, including the eyes. The thickness of the exoskeleton varies depending on the nature and function of the body part it covers. For example, the exoskeleton is thinner at the joints,

**arthropods** members of the phylum of invertebrates characterized by segmented bodies and jointed appendages such as antennae and legs

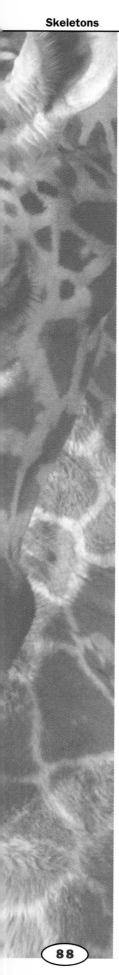

which require a degree of flexibility in order to bend. The chemical composition of the exoskeleton also differs depending on location and function.

Insects have three principle body segments—the head, thorax, and abodomen—and six segmented legs. Each segment is curved and hinged to its neighbor. Spiders have a thorax and abdomen and eight legs. Their head is fused to their thorax. Other arthropods, including scorpions, centipedes, and shrimp, may have more body segments than an insect and more legs than a spider.

Exoskeletons have two major disadvantages when compared to bony skeletons. Because they are composed of a blend of rigid, inorganic substances, they cannot expand as their owner grows. Arthropods must therefore shed their exoskeletons periodically through a process called molting. A newly **molted** animal is vulnerable to attack by predators before its new shell hardens. The shell hardens through a process similar to the tanning of leather.

**molted** the shedding of an exoskeleton as an animal grows so that a new, large exoskeleton can be secreted

The second disadvantage of exoskeletons is that physical contraints limit the size attainable by animals that have them. As the animal gets larger, the size of the exoskeleton required to cover its surface area would render the covering heavy and cumbersome.

## Hydrostatic Skeletons

**hydrostatic skeleton** a pressurized, fluid-filled skeleton

**Hydrostatic skeletons** are found mostly in aquatic organisms such as octopusses, jellyfishes, sea anemones, and tunicates. Although earthworms and velvetworms have hydrostatic skeletons as well. Bony skeletons and exoskeletons are made of relatively rigid substances, but hydrostatic skeletons contain no hard parts at all.

These soft, supporting structures have two components, a fluid-filled body cavity and a muscular body wall. Animals with hydrostatic skeletons move using the combined actions of these two features. They use the muscles of the body wall to squeeze fluid into other regions of the body cavity, allowing them to change shape. These shape changes allow the animal to extend parts of its body in the direction it wants to move and withdraw other parts from areas it is leaving.

One benefit that hydrostatic skeletons give to some soft-bodied organisms is an ability to take in important materials such as oxygen, water, and waste products through the skin. This eliminates the need for a separate transport system. It is beneficial for these animals not to have a separate transport system because transfer of these materials is a passive process, which means that energy does not need to be expended to take in oxygen, rid the body of waste products, and maintain the balance of water between the body and the environment. In addition, these skeletons are relatively light compared to bony skeletons and exoskeletons. This is beneficial because not as much muscle mass is required to move it. Hydrostatic skeletons work well in aquatic environments, but they would not be useful on land. They give little protection against drying out, and larger animals would be too flimsy to stand up on their own. SEE ALSO BONES; CARTILAGE; CHITIN; KERATIN.

*Judy P. Sheen*

# Social Animals

Sociability is a trait that applies to the ecology and behavior of a species and not to individual organisms. Social species are genetically inclined to group together and follow a particular set of rules defining interactions between individuals. Humans can be considered a social species because we tend to live in communities instead of segregating ourselves as individuals and dispersing to unoccupied territory. In many species, a family unit, meaning parents and their immediate dependent young, groups together and follows particular guidelines of interaction. However, this does not qualify as a society. A society must be composed of more individuals than are contained in a family unit. Even in typically antisocial species, individuals may temporarily unite to bear and raise young before re-dispersing.

Sociability in animals must be either permanent or semi-permanent, unlike family units. The species must also divide social responsibilities among individuals within the group. For example, one group of individuals, whether determined by age, gender, or body shape, must consistently perform a particular function. This requirement disqualifies animals that are merely nonaggressive with one another, but that do not partake in the formality of social structure.

## Definition

A rigorous definition of an animal society is: a group of animals belonging to the same species, and consisting of individuals beyond those in a family unit, who perform specific tasks, spend distinctly more time together, and interact much more within the group than with members of the same species outside of that group. A social animal is defined as any animal species that typically forms into societies.

Many ecologists are concerned with social behaviors, which are any behaviors specifically directed towards other members of the society. These can include cooperative, selfish, hurtful, or helpful behaviors. The sum of these behaviors determines the character of the society, such as its size and location, and the responsibilities of different societal members. For example, walruses live in coastal arctic regions, within herds containing up to several hundred individuals. Males within a herd are known to attack one another over disputes pertaining to female choice, territory, or food; however, entire herds have been known to come to the defense of a single member when placed in danger. Plants, fungi, and single-celled organisms are not considered social because their interactions are strictly dictated by physical and chemical needs. Thus they cannot behave, and without social behaviors a society is impossible.

## Why Species Form Societies

The tendency of a species to form into societies is considered to be caused by the influence of **natural selection**. During the process of evolution, individual animals of a particular species that formed into societies were more likely to survive than those animals that remained isolated from one another. This pressure to be social could be genetic in origin, meaning that high levels of kinship cause extended families to protect one another. It could also be an environmental pressure—a particularly harsh environment

**natural selection**
process by which organisms best suited to their environment are most likely to survive and reproduce

may force animals to depend on each other for greater support. A certain species that is observed to be social today may no longer be under selective pressure to form societies, but retains the trait of sociability for genetic reasons.

Some species may be social in certain environments and solitary in others. This can be linked to food abundance. For example, if food is readily available, individuals can group together for greater protection from predators. If food is scarce, they will remain as individuals so as to avoid the responsibility of supporting weaker individuals. This illustrates how societies can be both cooperative and competitive.

They are cooperative because the animals within a society must share their space and resources in order to ensure the survival of as many members as possible. However, they are competitive because each individual animal within a society is primarily concerned with its own survival. In the case of ample food, a strong animal may enlist the help of weaker animals. The stronger animal benefits from the extra labor and shares the surplus with the weaker animals who may not have survived on their own. This is cooperative behavior.

If the same society is deprived of food, however, the stronger animal no longer benefits from sharing its scarce supplies. It separates from the weaker animals, which are now forced to fend for themselves. This is competitive behavior. In less extreme situations, members of a healthy, normal society express both types of behavior at different times.

## Degrees of Sociability

Different species of animals may be social in greater or lesser degrees. The degree of sociability is loosely determined by the likelihood of forming a society, the degree of variation in societal size, the specificity of division of labor, the interdependence of members for survival, and possible **phenotypic** specializations, difference of body shape, related to societal rank. A non-social species may form into societies in extreme or abnormal situations. This is a common occurrence for captive **populations** of territorial animals that normally have large ranges in which they remain isolated. In zoos and research laboratories, these animals are artificially forced to live in close proximity to one another. They may display signs of cooperation that herald societal behavior, but this is most likely forced interaction that would not occur in the wild.

## The Sizes of Societies

The relative sizes of societies within the same species is a clue to understanding their degree of complexity. If members of the same species located in similar environments always form societies of approximately the same number of individuals, then that species may have complex societal laws dictating group size. Often, maximum group size is determined by local predation on the social species, when regional predators are better able to locate large groups than small groups of prey.

Some species maintain equivalent societal populations even in the absence of heavy predation, and these can be considered more socially complex. Highly complex societies divide tasks among societal members in a

**phenotypic** physical and physiological traits of an animal

**populations** groups of individuals of one species that live in the same geographic area

very precise manner. For example, some individuals may be responsible for hunting, some for reproducing, some for maintaining the society's living space, and some for raising the young. The clearer the distinction between the duties of societal members, the more complex that society is. In the most complex societies, individuals of a particular caste or rank almost never switch roles within their society. An animal may exchange roles only in situations where a particular caste that performs a particular function is severely depleted in numbers and needs to be augmented by new members.

In less complex societies, animals may have shifting duties and responsibilities. In such a society, a female may reproduce one year, but care for young the following year, and hunt the next year. The roles of individuals within the society continually shift based on the needs of the individual versus the needs of the group.

## Effects of Sociability

Occasionally, in species with a very high degree of task specificity, the **phenotype**, or set of body characteristics, of individuals is determined by their

These Chacma Baboons (*Papio ursinus*) from Zimbabwe form a highly social family group.

**phenotype** the physical and physiological traits of an animal

caste. This principle supercedes gender differences in body type. An animal's societal role may be so distinct that its body grows into a form that can best perform that role. Often, the role is easily determined by the appearance of the animal. This is a very rare phenomenon for vertebrates, although it is relatively common in insects such as honeybees and termites.

This **phenotypic variation** is often based on internal **hormones**, or external **pheromones**. Pheromones are hormones released into the environment that can alter the development, behavior, and appearance of individual animals. Pheromones have species-specific effects, meaning that they typically affect only animals within the same species. Type of food consumed, environmental conditions, and population ratios between the castes can affect the caste phenotype of individuals either directly or indirectly by altering the exposure of individuals to particular pheromones.

Sociability is a trait of many organisms, but two particular animals provide models of highly complex animal societies. Termites and naked mole rats come from entirely divergent evolutionary origins, yet they share a similar social organization. This raises the question for ecologists of why this social strategy is so effective that it arose multiple times in multiple branches of the evolutionary tree. Understanding the social interactions of these animals can contribute to the theoretical explanation of this organizational pattern.

The highest degree of sociability in animals is given its own classification term, **eusocial**. Eusocial animals must exhibit a reproductive division of labor, which means that many sterile individuals work to support those individuals capable of breeding. Furthermore, they must be organized into a system of discrete castes distinguishable by differences in skeletal or body characteristics, and these castes must contribute to care of the young. Finally, the generations must overlap in time, meaning that the young do not have time to mature before another group of young is produced. Termites and naked mole rats both conform to this definition.

## Termites

Termites live in carefully constructed structures entirely shut off from the light, usually within dead trees or wood products. Their social structure is relatively flexible compared to that of ants: In an ant colony, the sizes and body forms of individuals are fixed and constant at maturity, but termites change the shapes of their bodies depending on their role in the colony, even as adults.

Termite colonies may contain from several hundred to several million individuals. New colonies are founded annually during a short season when an existing colony produces sexual winged members that leave the nest by the thousands in a marriage flight. After about two and a half hours, the termites drop to the ground and purposefully break off their wings. The male and female dig a tunnel into a damp log for approximately two days and then mate within two weeks. Termites undergo incomplete development, which means that they resemble miniature, immature adults at hatching, and never pass through a grub stage of development. When the brood is old enough, they assume complete responsibility for feeding and caring for their parents, which are now called the royal couple, or the king and queen. Both

**phenotypic variation** differences in physical and physiological traits within a population

**pheromones** small, volatile chemicals that act as signals between animals that influence physiology or behavior

**hormones** chemical signals secreted by glands that travel through the bloodstream to regulate the body's activities

**eusocial** animals that show a true social organization

the king and queen grow much larger than other caste members, and the queen's abdomen swells with eggs until she becomes up to 20,000 times the size of other colony members.

Termites continuously exchange food and clean one another, probably to ensure cleanliness, but these behaviors also serve to cement the social interaction of the colony. Pheromones secreted by the royal couple and spread through saliva and food-sharing inhibit the development of more sexual forms in the colony, but the mechanisms underlying **differentiation** of other castes are unclear. They seem to develop when needed by the colony.

Termites are divided into more castes than most social insects, and the castes are far more varied. The basic division of labor includes the reproductive royal couple and sterile soldiers and workers. There are two possible types of soldiers, although any given species contains only one type. The mandibulate uses its large pincers to attack enemies, while the nasute excretes a sticky, poisonous substance from its elongated snout to immobilize enemies.

Soldiers only serve to defend the nest from enemies, and cannot feed themselves or perform any other function in the colony. Workers perform all of the construction tasks of the colony as well as feeding and cleaning the royal couple, eggs, and soldiers. Substitute sexual forms develop from juveniles when one or both of the royal couple dies. They battle amongst themselves until only one reproductively capable couple remains, and these two will develop into the new king and queen.

Like termites, every species of ant, most species of bees, and many species of wasp live in similarly-structured social colonies, with one reproductive pair and many sterile castes. This strategy has evolved independently many times in insects, which means that it must confer some general evolutionary advantage to the species. Perhaps by sequestering the only viable pair in the heart of a well-defended fortress, these insects are better able to ensure the safety of mating and survival.

## Naked Mole Rat

One species of vertebrate, the naked mole rat, has also adapted a colonial social structure. Naked mole rats are more closely related to voles than to either moles or rats. They mainly inhabit Somalia, Ethiopia, and Kenya, in sandy soil burrows that extend from near-surface level to several meters below the ground. These are the only known cold-blooded vertebrates. Their body temperatures are always equivalent to ambient temperature except when they are highly active.

These animals also differ from other mammals in that they live in a complex social environment. There is one breeding female per colony, and she can be considered the equivalent of a queen in the termite colony. She is longer-bodied and heavier than other members. Unlike the termite example, there is no single king to mate with the lead female; instead, any male in the colony is free to mate with the lead female, but there is little aggression or competition between these males. This female gives birth to one to six pups per litter, and approximately two litters per year, and these animals have a life span of twenty to forty years.

**differentiation** differences in structure and function of cells in multicellular organisms as the cells become specialized

As decomposers, termites are highly important members of their ecological communities. Termites feed on dead plant cell wall material, with their most prevalent form of food being wood. Termites themselves do not digest the cellulose; rather, the protozoans and bacteria that live symbiotically in the termites' gut break down the tough wood fibers. In some communities, termites can be responsible for recycling up to one third of the annual production of dead wood, thereby making many needed nutrients available to their environment.

Naked mole rats consume their own feces and that of other mole rats. They also roll their bodies in the feces to ensure that no nutrients are lost through fecal matter that has not been entirely digested, and to mark colony members with an identical signature scent.

Castes in the naked mole rats are partially determined by age and partially by body size. The leading female maintains her dominance in the colony through shoving and threatening non-breeders. All castes except for the breeding female engage in building, digging, and transporting food and soil, but the amount of work they perform is clearly delineated by their caste. Small-bodied juveniles are frequent workers, meaning that they carry the bulk of work activity of the colony. Slow-growing individuals remain frequent workers their entire lives, although larger, faster-growing individuals move into the infrequent workers caste, and eventually into the nonworkers caste.

The infrequent workers and nonworkers are most often colony-defenders. When the breeding female dies or is killed, the younger, smaller-bodied female caste members set upon one another. Their bodies begin to produce greater degrees of female hormones, and they fight until one dominant female remains alive. At this point, the colony recognizes their new breeding female, and colony life reestablishes itself.

Being social seems to confer many advantages to social animals, such as increased defense of breeding individuals, increased likelihood of survival, and increased stability of food stores and habitat. However, many animals are not social. An example is sloth bears, which hunt termites for food and come together only for the sake of reproduction. Explaining the reasons for the sloth bear's asocial behavior is not entirely possible without knowing its distinct evolutionary history. However, the bear's large body size and scarce food source may be reasons why colonial living is inappropriate for this species. SEE ALSO BEHAVIOR; DOMINANCE HIERARCHY; SOCIALITY.

*Rebecca M. Steinberg*

### Bibliography

Bennett, Nigel C., and Chris G. Faulkes. *African Mole-Rats: Ecology and Eusociality*, Cambridge and New York: Cambridge University Press, 2000.

Bonner, John Tyler. *The Evolution of Culture in Animals*. Princeton, NJ: Princeton University Press, 1980.

Caro, Timothy M. *Cheetahs of the Serengeti Plains: Group Living in an Asocial Species*. Chicago: University of Chicago Press, 1994.

Gadagkar, Raghavendra. *Survival Strategies: Cooperation and Conflict in Animal Societies*. Cambridge, MA: Harvard University Press, 1997.

Ito, Yoshiaki. *Behaviour and Social Evolution of Wasps: The Communal Aggregation Hypothesis*. Oxford and New York: Oxford University Press, 1993.

Wilson, Edward Osborne. *Sociobiology: The New Synthesis*, 25th anniversary ed. Cambridge, MA: Belknap Press of Harvard University Press, 2000.

# Sociality

Sociality is a genetically determined social behavior that dictates the social structure of particular groups of animals. Manifestations of sociality include living in close proximity, dividing tasks and responsibilities, and traveling together. Insects express sociality more frequently and to greater extremes than do most other animals, and as a result these animals have formed the archetype of all social animals.

Hives are the preferred "living arrangements" for eusocial bees. Various castes of bees will live harmoniously within the close living quarters of their hive.

Social insects live in every terrestrial **ecosystem** and form an important part of many food webs. **Eusocial** (extremely social) animals live in collective societal units referred to as hives or colonies. As adults eusocial animals serve their society in narrowly defined capacities know as "castes." Some examples of a caste are the worker, soldier, nurse, or reproductively capable king or queen.

In highly social animals all animals excepting the reproductive pair are sterile. In some animals the juvenile's role is predetermined, and it matures into the appropriate body size and shape. However, in other animals the caste of the mature adult is not decided until the animal has matured. In the latter case the individual's body size and abilities, and sometimes its age, will subsequently determine its caste position. There is evidence that some dinosaurs had complex social structures that caused them to herd and nest together. Although only one genus of vertebrate (the naked mole rat), is eusocial, many mammals, birds, reptiles, amphibians, and fish show social behaviors to a lesser degree.

Several theories have arisen to explain the evolution and conferred benefits of eusociality. One of these is "kin selection", suggested by W. D. Hamilton in his 1964 paper, "The Genetical Theory of Social Behavior." This theory suggests that social animals are often genetically very similar to each other. If one accepts that the evolutionary goal of a species is to en-

**ecosystem** a self-sustaining collection of organisms and their environment

**eusocial** animals that show a true social organization

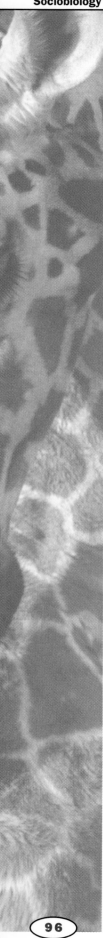

**genomes** the sum of all genes in a set of chromosomes

sure the survival of its DNA, then in animals with identical **genomes**, the survival and reproduction of individuals lose importance.

For example, if animal A and animal B have very different DNA, then each will desire to pass on its hereditary code to offspring because each code is unique. However, if A and B have very similar genomes, only one of them needs to reproduce for both of their genetic codes to be carried on to future generations. If the entire colony shares similar DNA, only two members of the colony are required to reproduce so as to pass on their genetic codes. This frees the remaining colony members from reproductive drive and allows them time and incentive to protect, feed, and house the reproductive couple. Thus the kinship of the animals ties them together.

The theory of kin selection is related to the idea of "inclusive fitness," which relates how a trait can pass from generation to generation directly, from parent to offspring, or indirectly through the help provided by individuals who possess the same trait as the parent. Wasps, bees, and ants have a genetic system in which all individuals are at least 50 percent identical, and this genetic similarity makes them a good candidate for kin selection and thus sociality. SEE ALSO Behavior; Dominance Hierarchy; Social Animals.

*Rebecca M. Steinberg*

**Bibliography**

Bourke, Andrew F. G., and Nigel R. Franks. *Social Evolution in Ants.* Princeton, NJ: Princeton University Press, 1995.

Frank, Steven A. *Foundations of Social Evolution.* Princeton, NJ: Princeton University Press, 1998.

Hamilton, W. D. "The Genetical Evolution of Social Behavior." *Journal of Theoretical Biology* 7 (1964): 1–52.

# Sociobiology

In 1975 Harvard University biologist Edward O. Wilson wrote a controversial book called *Sociobiology: The New Synthesis* in which he proposed to undertake "the systematic study of the biological basis of all social behavior" (Meachan 1998, pp. 110–113). Following a lifelong fascination with ants and their social structures, Wilson was interested in determining the degree to which genetic evolution influences cultural evolution. He applied the principles of evolution to the analysis of **behavioral** questions such as altruism, competition, cooperation, parasitism, dominance, **population** control, sex differences, and division of labor among social animals.

**behavioral** relating to actions or a series of actions as a response to stimuli

**population** a group of individuals of one species that live in the same geographic area

**natural selection** process by which organisms best suited to their environment are most likely to survive and reproduce

**genetics** the branch of biology that studies heredity

The theory of **natural selection** formulated by nineteenth-century British naturalist Charles Darwin would seem to indicate that the most successful animals would be those who act in their own self-interest. Yet clearly, social animals—societies of cooperating organisms—exist and often put the welfare of the group above their own. Wilson attempted to explain this using **genetics** and population models. He proposed that any organism—from the simplest to the most complex—exists as DNA's way of making more DNA. Therefore, a mother sacrificing herself for the sake of two or more offspring would overall benefit her DNA, as would the actions of a herd animal sacrificing itself for the sake of multiple related members.

Wilson attempted to apply sociobiology to human beings, saying that "We eat, sleep, build shelters, make love, fight and rear our young because, through the process of natural selection interacting with social influences, we developed genetic predispositions to behave in ways that ensured our survival as a species." This idea was violently rejected by fellow sociologists. It was taken to mean that everything humans did was predetermined genetically, although Wilson himself explained: "It is wrong to say that if a behavior is adaptive—that is evolutionarily advantageous—it cannot be conscious." He believed that while there is a biological basis for behavior such as nepotism, altruism, and status seeking, individuals within species are capable of immense variation. And further, he held that different species reacting to different environmental pressures tend toward particular forms of societies. Ant colonies, in other words, behave differently than elephant herds, and human groups behave like neither.

The questions raised by sociobiology have been incorporated into the larger field of behavioral **ecology**. These questions include: Which social arrangements best accommodate the general features of which species? What particular environmental conditions trigger different tendencies? Meanwhile, despite a superficial impression that nature is filled with constant competition, closer examination shows that cooperative behavior occurs throughout **ecosystems** and is far more stable and beneficial to all the species involved. SEE ALSO BEHAVIOR; DOMINANCE HIERARCHY; SOCIAL ANIMALS; SOCIALITY.

*Nancy Weaver*

**ecology** the study of how organisms interact with their environment

**ecosystems** self-sustaining collections of organisms and their environments

**Bibliography**

*Concise Dictionary of Scientific Biography*. New York: Scribner. 2000.

Meachan, Dyan. "Please Pass the Ants." *Forbes* 164, no. 6 (1998): pp. 110–113.

**Sponges** *See Porifera.*

# Spontaneous Generation

From the seventeenth century, through the Middle Ages, and until the late nineteenth century, it was generally accepted that some organisms originated directly from nonliving matter. Such "spontaneous generation" appeared to start in decaying food, urine, and manure because worms or maggots could be seen hatching there after a few days. It was also believed that animals that lived in mud, such as frogs and salamanders, were generated by the mud in which they lived. Additionally, there were the widely held misconceptions that rats were spontaneously generated in piles of garbage or created from magical recipes. One seventeenth-century recipe even called for the creation of mice from sweaty underwear and wheat husks placed together in a jar for twenty-one days. Although such a concept may seem ludicrous today, it was congruous with other cultural and religious beliefs of the time.

Francesco Redi, an Italian physician, naturalist, and poet, first challenged the idea of spontaneous generation in 1668. At that time, it was widely held that maggots arose spontaneously in rotting meat. Redi did not believe this.

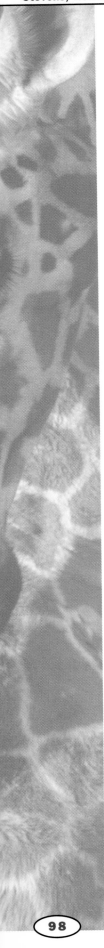

He hypothesized that maggots developed from eggs laid by flies. To test his hypothesis, he set out meat in a variety of jars, some open to the air, some sealed completely, and some covered with gauze. As Redi had expected, maggots appeared only in the jars in which the flies could reach the meat and lay their eggs.

Unfortunately, many people who were told or read about these experiments did not believe the results, so if they still wanted to believe in spontaneous generation, they did. Even Redi continued to believe that it occurred under some circumstances and cited the example of grubs developing in oak trees. The invention of the microscope during this time only seemed to further fuel this belief, as microscopy revealed a whole new world of microorganisms that appeared to arise spontaneously.

The debate over spontaneous generation continued for centuries. In the mid-eighteenth century, two other well-documented experiments—one by John Needham, an English naturalist, and the other by Lazzaro Spallanzani, an Italian physiologist—were attempted but were considered by proponents of spontaneous generation to be unpersuasive.

The idea of spontaneous generation was finally laid to rest in 1859 by the French chemist, Louis Pasteur. The French Academy of Sciences sponsored a competition for the greatest experiment that could either prove or disprove spontaneous generation. Pasteur devised a winning experiment where he boiled broth in a flask, heated the neck of the flask in a flame until it became pliable, and bent it into the shape of an "S." With this configuration, air could enter the flask, but airborne microorganisms could not, they would settle by gravity in the neck of the flask. As Pasteur had expected, no microorganisms grew. However, when he tilted the flask so that airborne particles could enter, the broth rapidly became cloudy with life. Pasteur had both refuted the theory of spontaneous generation and demonstrated that microorganisms are everywhere, including the air. SEE ALSO BIOLOGICAL EVOLUTION.

*Stephanie A. Lanoue*

**Bibliography**

Krebs, Robert. *Scientific Development and Misconceptions Throughout the Ages.* Westport, CT: Greenwood Press, 1999.

*Random House Dictionary of Scientists.* New York: Random House, 1997.

## Starfishes   *See Echinodermata.*

# Stevens, Nettie Maria

### *American biologist*
### *1861–1912*

Nettie Maria Stevens was a prominent biologist who discovered that the sex of an organism was determined by a specific chromosome. Although her research career spanned less than a decade (1903–1912), she published forty papers and became one of the first American women to achieve recognition for her contributions to scientific research.

Stevens was born July 7, 1861 in Cavendish, Vermont. She graduated from the Westford Academy in 1880 and enrolled at the Westfield Normal School, a educator college founded to cerify teachers, in 1881. She received the highest scores on the college's entrance exams of any student in her class. Clearly an outstanding student, she earned the four-year teacher certification in just two years.

In 1883, Stevens graduated from Westfield Normal School with the highest academic scores in her class. After graduation she took a job as a librarian at a high school in Lebanon, New Hampshire. In 1896, at the age of 35, she transferred to Stanford University in California, where she earned a B.A. degree in General Biology in 1899 and an M.A. in **physiology** in 1900. She began her doctoral studies at Bryn Mawr which included a year of study (1901–1902) at the Zoological Station in Naples, Italy, and at the Zoological Institute of the University of Würzburg, Germany.

Stevens received a Ph.D. in Morphology from Bryn Mawr in 1903 and remained at the college as a research fellow in biology the following year. She was an associate in experimental morphology at Bryn Mawr from 1905 until her death from breast cancer on May 4, 1912, in Baltimore, Maryland at the age of 51.

Stevens' earliest field of research was the morphology and **taxonomy** of ciliate **protozoa**. One of her major papers in that field was written in 1904 with zoologist and geneticist Thomas Hunt Morgan, who won the Nobel Prize in 1933 for his work. Stevens' investigations into regeneration led her to a study of **differentiation** in embryos and then to a study of **chromosomes**. Steven's research showed that very young embryonic cells could not fully regenerate. In 1905 Stevens published a paper in which she announced her landmark finding that the chromosomes known as X and Y were responsible for the determination of the sex of an individual.

*Stephanie A. Lanoue*

Nettie Maria Stevens was one of the first American women to have her scientific achievements recognized.

**physiology** study of the normal function of living things or their parts

**taxonomy** the science of classifying living organisms

**protozoa** a phylum of single-celled eukaryotes

**differentiation** differences in structure and function of cells in multicellular organisms as the cells become specialized

**chromosomes** structures in the cell that carry genetic information

**Bibliography**

Morgan, Thomas Hunt. "The Scientific Work of Miss N. M. Stevens." *Science* 36, no. 928 (1912):21–23.

Ogilvie, Marilyn Bailey, and Clifford J. Choquette. "Nettie Maria Stevens (1861–1912): Her Life and Contributions to Cytogenetics," *Proceedings of the American Philosophical Society*, 125, no. 4 (1981):10–12.

Shearer, Benjamin F., and Barbara S. Shearer. *Notable Women in Life Sciences.* Westport, CT: Greenwood Press, 1996.

## Surface-to-Volume Ratio *See Body Size and Scaling.*

# Sustainable Agriculture

Thanks to a complex international system of production, processing, shipping, and marketing, people today can eat a vast selection of out-of-season and out-of-region fruits and vegetables year round. In addition, modern farming practices have provided an abundant, not just a varied, food supply. However, the availability of abundant and varied food relies on energy-intensive,

nonrenewable resources such as fossil fuels, and many practices associated with agribusiness have had a detrimental impact on the environment.

## Agribusiness

Agribusiness looks at farms as factories to be run as profitably and efficiently as possible. "How much, how fast" replaces the old values of carrying capacity (how much the land can yield without being depleted) and a season of lying fallow. Monocropping (fields used for only one crop, the same crop year after year, commonly wheat, soybeans or cotton) and increased field size do away with **biodiversity** and hedgerows, and thus with fertility, pollinators, and resistance to insects and disease.

**Effects of fertilizers.** Millions of tons of chemical fertilizers applied to fields destroy microorganisms that are vital to the health of the soil. The intensive use of herbicides and **pesticides** kills pollinating insects that are essential for crop production. Manure and crop residues, once valuable sources of soil nutrition, are no longer tilled in and have become polluting waste products themselves, to be burned or dumped. Underground aquifers (natural water reservoirs) that took thousands of years to fill are pumped dry to irrigate fields in semi-arid regions.

Additionally, ground water is full of dissolved mineral salts. As the water evaporates it leaves a salt concentration buildup that is poisoning the soil. Intensive plowing opens up hundreds of thousands of acres of topsoil to erosion by wind and rain, filling the air with dust or silting up waterways.

**Loss of biodiversity.** As more and more wild land is converted to growing one particular crop on a massive scale, genetic biodiversity diminishes. Rain forests are bulldozed to provide pastures for cheap beef. Massive feedlot operations are susceptible to catastrophic diseases, and they pollute drinking water supplies with their tons of confined manure.

**Economic consequences.** Agribusiness also affects the economy. Family farmers are increasingly replaced by corporate managers, and the price of farm equipment and capital outlay soars. Profits from production may go to a large company based far away from the actual farm and never enter the local economy. In addition, large-scale food production and distribution have become vulnerable to the vagaries of international politics and the stock market, and any breakdown in the network places everyone at risk. The price of oil, interest rates, trucking fees, politics and the weather all affect the availability and price of food. In countries where there is economic or military chaos, even though there might be plenty of food on farms or at food aid centers, the breakdown in the complex distribution system results in famine.

## Sustainable Agriculture

Sustainable agriculture is the practice of working in concert with nature to replenish the soil in order to assure a secure, affordable food supply without depleting natural resources or disrupting the cycles of life. Proponents of sustainable agriculture suggest we can reverse the damage done by agribusiness. They believe that a dependable long-term food supply must rely on the protection of resources—seeds, food species, soil, breeding stock,

**biodiversity** the variety of organisms found in an ecosystem

**pesticides** substances that control the spread of harmful or destructive organisms

A farmer spreads fertilizer on his fields in northeastern Pennsylvania.

and the water supply, as well as the farmer who knows and cares for a particular piece of land and the community with which the farmer is interdependent. Sustainable agriculture promotes regional and local small-scale farms that rely on the interplay of crops and livestock to replenish the soil and control erosion. The aim of a healthy farm is to produce as many kinds of plants and animals as it reasonably can. Ordered diversity is the practice of maintaining many kinds of plants and animals together to complement one another.

**Cover cropping.** The practice of planting a noncommercial crop on fields to increase fertility, conserve soil moisture, keep topsoil from eroding or blowing away and encourage soil microorganisms is called cover cropping. Soil fertility, which is the major capital of any farm, can be largely maintained within the farm itself by this method and by plowing back in manure and other organic wastes. Food grown for local consumption is more fresh, can be harvested when ripe, and uses less energy to get to market. Farm stands and local farmers' markets provide more money for the farmer and higher quality, lower-price food for the consumer.

**Diversity of methods.** Sustainable agriculture embraces diversity of method and scale, looking for what is appropriate to a given location. One example is urban homesteading, in which thousands of vacant inner-city lots can be used to grow neighborhood gardens. Renewable energy sources, such as passive-solar greenhouses or windmills, are encouraged. Sustainable agriculture advocates organic solutions to pest control such as crop rotation, the introduction and maintenance of beneficial insects, and intercropping (growing more than one kind of crop on the same land in the same growing season). All these methods discourage insect infestations, thus reducing the amount of pesticides in the environment.

Agriculture cannot survive for long at the expense of the natural systems that support it. And a culture cannot survive at the expense of its agriculture. SEE ALSO FARMING.

*Nancy Weaver*

**Bibliography**

Berry, Wendell. *The Unsettling of America: Culture and Agriculture.* San Francisco: Sierra Club Books, 1977.

Organic Farming Research Foundation. *Winter 2001 Information Bulletin*, no. 9. Erica Walz, ed. Santa Cruz, CA: Organic Farming Research Foundation, 2001.

# Systematist

Systematics is the field of biology that deals with the diversity of life. It is the study of organisms living today and in the past, and of the relationships among these organisms. Systematics includes the areas of taxonomy and phylogenetics. Taxonomy is the naming, describing, and classification of all living organisms and fossils. Phylogenetics is the study of evolutionary relationships among organisms. It is also the study of the physical and environmental settings in which evolutionary changes occurred.

**biogeography** the study of the distribution of animals over an area

Systematics is also an essential part of other fields such as **biogeography** (the mapping of where species occur), ecology (the study of the habitats and environmental factors that control where species occur), conservation biology, and the management of biological resources.

Systematists collect plants and animals, study them, and group them according to patterns of variation. Systematists study plants and animals in nature, laboratories, and museums. Some study the scientific basis of classifications so they can better understand evolution. Others study ever-changing aspects of nature, such as the processes that lead to new species or the ways that species interact. Other systematists study the human impact on the environment and on other species. Some systematists screen plants for compounds that can be used for drugs. Others are involved in controlling pests and diseases among plant and animal crops. Many systematists have teaching careers. They may work in colleges and universities. They teach classes, teach students how to conduct research, and conduct their own research in their particular area of interest. An important part of the research is writing up the results for publication.

Systematists are employed mostly by universities, museums, federal and state agencies, zoos, private industries, and botanical gardens. Universities with large plant or animal collections often hire systematists as curators to maintain the collections and conduct research on them. Federal and state agencies employ systematists in many fields including public health, agriculture, wildlife management, and forestry. Industries that employ systematists include agricultural processors, pharmaceutical companies, oil companies, and commercial suppliers of plants and animals. Most jobs in government and industry center on taxonomy and **ecology**, rather than evolution issues.

**ecology** study of how organisms interact with their environment

At the high-school level, persons interested in becoming systematists should study math, chemistry, physics, biology, geology, English, writing, and computer studies. Although there are career opportunities for systematists with bachelor's degrees, most professionals have either a master's or doctoral (Ph.D.) degree. Undergraduate degrees can be obtained in biology, **botany**, or zoology. Graduate students focus specifically on systematics. They study taxonomy, **population** biology, genetics, evolution, ecology, biogeography, chemistry, computers, and statistics. SEE ALSO MORPHOLOGY.

*Denise Prendergast*

**botany** the scientific study of plants

**population** a group of individuals of one species that live in the same geographic area

### Bibliography

Keeton, William T., James L. Gould, and Carol Grant Gould. *Biological Science.* 5th ed. New York: W.W. Norton & Company, Inc., 1993.

### Internet Resources

American Society of Plant Taxonomists: <http://www.csdl.tamu.edu/FLORA/aspt/asptcar1.htm>.

Society for Integrative and Comparative Biology: <http://www.sicb.org/>.

## Tapeworms *See Cestoda.*

## Taxonomist

Taxonomy is the naming, describing, and classification of all living organisms and fossils. It is an important part of systematics, the field of biology that deals with the diversity of life. There are three main schools of taxonomy:

- Phenetic taxonomy—which classifies organisms based on overall form and structure or genetic similarity, also called numerical taxonomy;

- Cladistic taxonomy—which classifies organisms strictly based on branching points, focuses on shared, relatively recent (not of ancient origin) characteristics that are common to the species being studied, also called **phylogenetic** taxonomy; and

- Evolutionary taxonomy—which classifies organisms based on a combination of branching and divergence, also called traditional taxonomy.

**phylogenetic** relating to the evolutionary history of species or group of related species

Taxonomists collect plants and animals, study them, and group them according to patterns of variation. Taxonomists study plants and animals in nature, laboratories, and museums. Several million species of animals and over 325,000 species of plants are presently known. It is estimated that between several million and 30 million species await discovery. Animal species are included in the International Code of Zoological Nomenclature, a document that also contains the rules for assigning scientific names to animals.

Some taxonomists study the scientific basis of classifications so they can better understand evolution. Others study the ever-changing aspects of nature, such as the processes that lead to new species or the ways that species interact. Other taxonomists study the human impacts on the environment and on other species. Some taxonomists screen plants for compounds that can be used for drugs. Others are involved in controlling pests and diseases

among plant and animal crops. Many taxonomists have teaching careers. They may work in colleges and universities. They teach classes, teach students how to conduct research, and conduct their own research in their particular area of interest. An important part of the research is writing up the results for publication.

Taxonomists are generally employed by universities, museums, federal and state agencies, zoos, private industries, and botanical gardens. Universities with large plant or animal collections often hire taxonomists as curators to maintain the collections and conduct research on them. Federal and state agencies employ taxonomists in many fields including public health, agriculture, wildlife management, and forestry. Industries that employ taxonomists include agricultural processors, pharmaceutical companies, oil companies, and commercial suppliers of plants and animals.

At the high-school level, persons interested in becoming taxonomists should study mathematics, chemistry, physics, biology, geology, English, writing, and computer studies. Although there are career opportunities for taxonomists with bachelor's degrees, most professionals have either a master's or doctoral (Ph.D.) degree. Undergraduate degrees can be obtained in biology, botany, or zoology. Graduate students focus specifically on taxonomy. They study population biology, **genetics**, evolution, ecology, biogeography, chemistry, computers, and statistics. SEE ALSO TAXONOMY.

*Denise Prendergast*

**genetics** the branch of biology that studies heredity

### Bibliography
Keeton, William T., James L. Gould, and Carol Grant Gould. *Biological Science.* 5th ed. New York: W.W. Norton & Company, Inc., 1993.

### Internet Resources
American Society of Plant Taxonomists: <http://www.csdl.tamu.edu/FLORA/aspt/asptcar1.htm>.

Society for Integrative and Comparative Biology: <http://www.sicb.org/>.

## Taxonomy

Taxonomy is the practice of classifying and naming biological organisms and groups of biological organisms. Modern taxonomy originated in the eighteenth century with the work of a Swedish botanist, Carolus Linnaeus, who developed an organized hierarchical classification system that could be applied to all biological organisms. This system is still referred to as Linnaean taxonomy. Under the Linnaean system, all organisms are classified hierarchically into a series of groups from most inclusive to least inclusive. Each species belongs successively to a kingdom, a phylum, a class, an order, a family, a genus, and a species. Sometimes additional categories are used, such as subphylum, superorder, superfamily, subfamily, etc.

One important aspect of taxonomy is the description and naming of biological species. Since the time of Linnaeus, species have been named using what is called binomial nomenclature, in which each species has a two-word name that includes a genus name and a species name. Species names are universal, that is, the same names are used by scientists worldwide. In addition, no two species can have the same name. If more than one name has been applied to a given species, the oldest name generally takes precedence. There are a few rules that govern the choice of species names. The entire binomial name must be in Latin or must be Latinized, with the genus name capitalized and the species name in lowercase. The binomial species name is always italicized or underlined in print.

As an example of taxonomic classification, the domesticated dog belongs to the kingdom Animalia (multicellular animals), the phylum Chordata (multicellular animals possessing a nerve cord), the subphylum Vertebrata (having a backbone), the class Mammalia (mammals), the order Carnivora (which includes carnivores such as dogs, cats, weasels, and hyenas), the family Canidae (including dogs, wolves, foxes), the genus *Canis*, and the species *familiaris*. The binomial species name for the domesticated dog is therefore *Canis familiaris*, where *Canis* designates the genus, and *familiaris* designates the species.

In Linnaeus's time, classification was based primarily on similarities between species, particularly anatomical similarities. Modern classification systems remain hierarchical, but use evolutionary relationships, rather than similarity among organisms, as the basis of classification. In order to achieve this, taxonomists try to describe supraspecific **taxa** that are **monophyletic** —that is, they attempt to create supraspecific taxa that include all the species that are descended from a common ancestor. Monophyletic groups are also known as **clades**. Many currently recognized groups are not monophyletic. For example, the group Reptilia, as it is traditionally conceived, is not monophyletic. Traditionally, Reptilia has included only turtles, lizards, snakes, and crocodiles. However, an analysis of the evolutionary relationships among different vertebrate groups reveals that the common ancestor of the traditionally recognized reptiles also gave rise to birds. Therefore, to make the group Reptilia monophyletic, birds should be included as reptiles.

The study of evolutionary relationships among organisms, which makes up the field of systematics, is a necessary first step in the creation of a modern taxonomy. As scientists study evolutionary relationships among organisms, they make changes in taxonomy in order to eliminate nonmonophyletic groups. Consequently, taxonomy is constantly changing as new knowledge is gained. SEE ALSO CLASSIFICATION SYSTEMS; CAROLUS LINNAEUS; PHYLOGENETIC RELATIONSHIPS OF MAJOR GROUPS.

*Jennifer Yeh*

**taxa** named taxonomic units at any given level

**monophyletic** a taxon that derived from a single ancestral species that gave rise to no other species in any other taxa

**clades** a branching diagram that shows evolutionary relationships of organisms

**Bibliography**

Futuyma, Douglas J. *Evolutionary Biology.* Sunderland, MA: Sinauer Associates, 1998.

Ridley, Mark. *Evolution.* Boston: Blackwell Scientific, 1993.

# Territoriality

A pattern of animal behavior, territoriality implies a fixed area (or territory) from which intruders are excluded by the owner through a combination of advertising, threatening, and attacking behaviors. It is important to distinguish between a territory and a **home range**, because the appearance of an

**home range** the area where an animal lives and eats

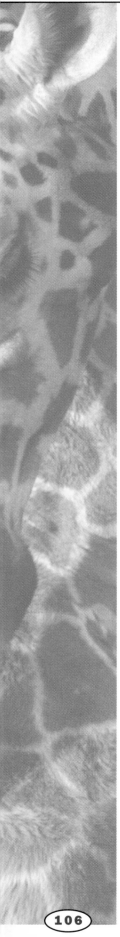

These elk rutting bulls fight in a field in Yellowstone National Park.

outsider will elicit different reactions from the animal that lives in, or frequents, the area. Unlike an animal's marked territory, its home range is an area in which the individual roams about but the individual rarely defends it against other animals. There are some species, such as the breeding song sparrow, whose territory and home range are one and the same. In the majority of species, however, the territory tends to be smaller than the home range or the two areas overlap so that only part of the home range is defended as territory.

A question most often asked regarding the defense of a territory is: Why do owners usually win? In weighing the benefits versus the costs of defense, the owner is more likely to escalate the battle than the intruder because the owner has already invested in the area of conflict. The land is worth more to the owner not only because of familiarity with the area but also because the individual may be fighting to maintain exclusive access to the resources found on the territory. These resources may include one or more of the following: food, water, nest sites, and potential/current mates.

There are some costs of defense that the owner must assess as well. There may be time loss from other activities, such as foraging for food or mating. There is also an energy cost in defending an area. The signaling activities can be energetically expensive be they through continual chattering (e.g., squirrels), proclamations through singing (e.g., songbirds), or leaving scent marks at different points in and around the territory (e.g., bears). Further energy is expended in patrolling the perimeters and chasing off any animal that is getting too close to the boundaries or has crossed over into the territory. Finally, there is the risk of injury in battle with any intruder as well as a risk of predation as the owner focuses more on guarding and is therefore less guarded against attack.

There are three categories of territories: breeding, feeding, and all-purpose. The breeding territory is relatively small. It usually contains only a nesting or mating site. This type of territory is most characteristic of colonially nesting species that cluster nests at limited safe sites, such as in lekking or chorusing species where the males aggregate to attract females. The feeding territory tends to be larger than the breeding territory because it must contain sufficient food to support the owner of the territory and any mate or offspring that may also be residing there. Defense of the feeding territory is greatest during the nonbreeding season, because the individual's attention is more focused on the territory. The owner also becomes more vigilant during times when food is scarce. The cost of defense is worthwhile to the owner as it ensures the individual exclusive access of the area's resources. Finally, the all-purpose territory is generally the largest as it includes aspects of both breeding and feeding territories.

For all three of these types of territories, there are usually adjacent territories contiguous with an individual's proscribed area. The owner of a particular territory may have as few as two and as many as six neighbors with whom it shares common boundaries. The network of these contiguous territories is known collectively as a "neighborhood."

Territorial defense is generally employed only against animals of the same species, because animals of a different species will often inhabit a different **niche** within the same territory. In this manner, different species can coexist in the same area and not impinge on each other's food resources. Furthermore, there is no threat of the other animal stealing the owner's mate.

**niche** how an organism uses the biotic and abiotic resources of its environment

With the hierarchies that are found within communities, the territorial system comes into play as the stronger, more aggressive animal generally wins the better territory and maintains it against others. These systems have effects not only on an individual basis but also at the **population** level as well. If resources were allocated "fairly" to each member of a community or a species, then there actually may not be enough to sustain any one individual. (If resources become scarce because of a fire or drought, the individual may expand its territory in order to find sustenance.) This kind of division of territories, and thereby resources, would lead to population crashes. While it may be difficult to establish one's own territory, that kind of competition is necessary in order for the species to survive. SEE ALSO FORAGING STRATEGIES; HOME RANGE.

**population** a group of individuals of one species that live in the same geographic area

*Danielle Schnur*

**Bibliography**

Alcock, John. *Animal Behavior: An Evolutionary Approach*, 4th ed. Sunderland, MA: Sinauer, 1989.

Bradbury, Jack W., and Sandra L. Vehrencamp. *Principles of Animal Communication*. Sunderland, MA: Sinauer, 1998.

Campbell, Neil A. *Biology*, 3rd ed. Berkeley, CA: Benjamin/Cummings, 1993.

# Tertiary

The Tertiary era, from 65 to 2 million years ago, consists of six epochs: the Paleocene, Eocene, Oligocene, Miocene, and Pliocene, which represent

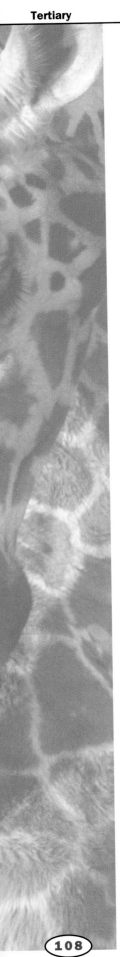

chapters in the story of the mammal's rise to dominance of land and oceans. The Tertiary follows the great Cretaceous extinction in which the dinosaurs, who had dominated the terrestrial food chain for hundreds of millions of years, inexplicably vanished, leaving only a few reptiles and mammal-like creatures as survivors.

The ancestors of the mammals, the therapsids, had been evolving into a broad range of ecological niches since the Permian–Triassic periods, some 260 million years ago. During the Mesozoic reign of dinosaurs, these mammals had dwindled almost into nonexistence, a few rat-sized species eking out a **nocturnal** insectivorous living, staying out of the way of predators.

**nocturnal** active at night

This long period of trying to avoid being eaten may actually have produced the very features that later allowed mammals to spread across the entire planet. Smaller animals had a greater need for maneuverability, selecting for skeletal changes toward speed and flexibility of joints and spine. Smaller mammals need proportionately greater energy to maintain a constant body temperature, thereby selecting for more efficient teeth and jaws as well as digestive systems. And what may have seemed like their greatest drawback, the birth of helpless young who need a period of parental care, actually produced offspring who were uniquely flexible in their behavior patterns and able to be taught by their parents.

In the Tertiary with the dinosaurs gone, mammals along with birds underwent a cycle of massive evolutionary expansion into the greatest range of shapes and sizes to ever populate Earth.

**plate tectonics** the theory that the Earth's surface is divided into plates that move

The story of evolution parallels that of geography. During the Permian period (250 million years ago) the supercontinent of Pangaea allowed for migration of plants and animals across the whole of Earth. When Pangaea, driven by the forces of **plate tectonics**, began to break up into separate continents, each chunk of land took with it a random cargo of the original inhabitants. Separation breeds diversity and all of the earliest archetypes (orginal ancestors of a group of animals), grazers, browsers, carnivores, **insectivores**, and canopy dwellers were free to evolve in wildly different ways.

**insectivores** animals who eats insects

In the first epoch of the Tertiary, the Paleocene (65–55 million years ago), mammals still consisted of survivors from the Cretaceous, including the **monotremes**, primitive egg-laying mammals.

**monotremes** egg-laying mammals such as the platypus and echidna

Condylarths, the ancestors of the **ungulates**, or hoofed animals, were widely present in the Paleocene. This group included carnivores and **scavengers**, as well as more common **herbivores**. Some rodent-like early primates also appeared in the Northern Hemisphere during this epoch.

**ungulates** animals with hooves

**scavengers** animals that feed on the remains of animals that it did not kill

In the Paleocene seas, sharks became the most abundant fishes, while **gastropods** and **bivalves** replaced the once-dominant ammonites.

**herbivores** animals who eat plants only

By the Eocene, also known as the "dawn of early life" (55–39 million years ago), Pangaea had begun to break apart. Australia had split off, carrying a load of marsupials, mammals who give birth to immature young who then crawl into a pouch (marsupium) in which they suckle and grow. Freed from competition with placental mammals, the marsupials diversified into every ecological niche across the whole of the Australian continent. Limestones in northern Queensland reveal a tropical rainforest of marsupials for every niche.

**gastropods** mollusks that are commonly known as snails

**bivalves** mollusks that have two shells

| Era | Period | Epoch | Million Years Before Present |
|---|---|---|---|
| Cenozoic | Quaternary | Holocene | 0.01 |
| | | Pleistocene | 1.6 |
| | Tertiary | Pliocene | 5.3 |
| | | Miocene | 24 |
| | | Oligocene | 37 |
| | | Eocene | 58 |
| | | Paleocene | 66 |

The Tertiary period and surrounding time periods.

Eocene seas chronicle the momentous return of the first mammals to the oceans they had emerged from several hundred million years earlier. The legs of the first whales began to change to flippers and increase in size, thanks to the new weightless environment.

In the Oligocene (39–22 million years ago), the Antarctic ice cap was beginning to form, provoking a marked cooling effect and a pattern of seasonal fluctuations. These changes apparently favored **homeothermic**, or warm-blooded animals, because the turtles, lizards, and crocodiles of the time did not undergo the explosion of evolution (cycles of immense activity and then decline in evolution) that the mammals underwent.

**homeothermic** describes animals able to maintain their body temperatures

By the Miocene (22–5 million years ago), a dryer, warmer **climate** again produced changes in vegetation which rippled through the world of herbivores and predators. Teeth patterns of Miocene fossils suggest that the vast **deciduous** forests and their leaf-browsing inhabitants were being replaced by vast grasslands and grazing animals. These early **ruminants** (cud-chewing animals) included several types of deer and the first horses. Predation tends to shape evolution, and the new open plains encouraged longer, swifter legs in horses or burrowing capabilities in smaller animals closer to the ground. Condylarths and creodonts, the flesh-eating ungulates, had begun to decline, replaced by other orders of carnivores that were faster, and had sharper teeth and claws.

**climate** long-term weather patterns for a particular region

**deciduous** having leaves that fall off at the end of the growing season

**ruminants** plant eating animals with a multi-compartment stomach such as cows and sheep

By the Pliocene, (5–2 million years ago) the continents had shifted into more or less their present-day locations. The **isthmus** of Panama had arisen to reconnect North and South America, allowing animals that had developed independently for millions of years to mingle. The two-way traffic across the isthmus sent the ponderous sloths and glyptodonts (giant armadillos) north. Highly evolved predators (such as sabre-toothed cats) traveled south, leaving numerous extinctions of South American marsupials in their wake. The isthmus also separated the ocean into two, Atlantic and Pacific, causing **differentiations** in marine species. SEE ALSO GEOLOGIC TIME SCALE.

**isthmus** a narrow strip of land

**differentiation** differences in structure and function of cells in multicellular organisms as the cells become specialized

*Nancy Weaver*

**Bibliography**

Asimov, Isaac. *Life and Time*. Garden City, NY: Doubleday & Company, 1978.

Fortey, Richard. *Fossils: The Key to the Past*. Cambridge, MA: Harvard University Press, 1991.

———. *Life: A Natural History of the First Four Billion Years of Life on Earth*. New York: Viking Press, 1998.

Friday, Adrian, and David S. Ingram, eds. *The Cambridge Encyclopedia of Life Sciences*. London: Cambridge University, 1985.

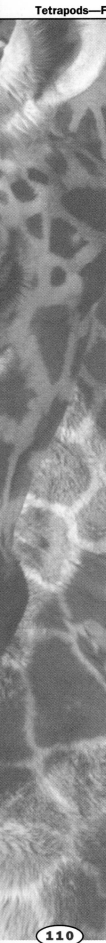

Gould, Stephen Jay, ed. *The Book of Life.* New York: W. W. Norton & Company, 1993.

Lambert, David. *The Field Guide to Prehistoric Life.* New York: Facts on File, 1985.

McLoughlan, John C. *Synapsida: A New Look Into the Origin of Mammals.* New York: Viking Press, 1980.

Steele, Rodney, and Anthony Harvey, eds. *The Encyclopedia of Prehistoric Life.* New York: McGraw Hill, 1979.

Wade, Nicholas, ed. *The Science Times Book of Fossils and Evolution.* New York: The Lyons Press, 1998.

# Tetrapods—From Water to Land

**vertebrates** animals with a backbone

Tetrapods—including the modern forms of amphibians, reptiles, birds, and mammals—are loosely defined as **vertebrates** with four feet, or limbs. Many species we see today, like the snakes or whales, may not appear to be tetrapods, but their lack of well-developed limbs is a secondary adaptation to their habitat. This means that they originally had four limbs, but lost them as they adapted to a certain style of living. In the **fossil record** scientists often see intermediate forms which has reduced limbs. In modern skeletons of these animals you can often see vestiges of appendages that indicate that they are, indeed, tetrapods.

**fossil record** a collection of all known fossils

## Life on Land

The appearance of tetrapods on land signaled one of the most hazardous and important evolutionary events in the history of animals. Life began in water. The body systems of early organisms were adapted to a mode of life in which water provided **buoyancy** against gravity. Desiccation, or drying out, was not a problem for animals whose bodies were constantly bathed in fluid. Movement from place to place usually required little energy and often involved simply floating along with the current. Reproduction was easier when sperm and eggs could be released into the water for **fertilization**. So the transition from living in the ocean to living on land required that ancestral vertebrates (who gave rise to the tetrapods) have physical traits that would helped them make this shift.

**buoyancy** the tendency of a body to float when submerged in a liquid

**fertilization** the fusion of male and female gametes

Tetrapods were not the first animals to make the move to land. Around 400 million years ago, primitive **arthropods** quickly followed the invasion of the first land plants, such as the mosses and liverworts, the first organisms to establish a foothold in the drier, but still moist, habitats, such as shorelines streams, and marshes. Both plants and insects continued to evolve and invade increasingly arid and varied habitats. They provided an important and diverse potential food base for the future land vertebrates.

**arthropods** members of the phylum of invertebrates characterized by segmented bodies and jointed appendages such as antennae and legs

## Fins and Legs

**aquatic** living in water

Paleontologists believe that only 50 million years after the first plants left their **aquatic** environments, two conditions existed that paved the way for the first tetrapods. First, competition for food in the oceans was extremely fierce. Second, a group of bony fishes called the lobe-finned (or sarcopterygian) fishes had developed the physical characteristics necessary for the transition from water to land.

In particular, one group of lobed-fin fishes, the Rhipidistians, had a general **body plan** that was very similar to and can be traced back to early amphibians. Rhipidistians had pairs of front and back fins with sturdy bones, instead of **cartilaginous** rods, for resting their bulky bodies on the sea floor. These special fins were strengthened by a particular arrangement of the bones that resembled the structure of tetrapods in many ways. Large single, heavy bones in the fins were located nearest to the body and attached to a central girdle, like the pelvis.

When the land animals began to use the fins for standing and walking, the girdle bones had to become stronger to support the body on the legs. These bones eventually became the femur (upper leg bone) and humerus (upper arm bone) of the tetrapods. Pairs of bones (the radius and ulna of the arm, and the tibia and fibula of the leg) were attached to the femur and humerus, followed by a series of smaller bones that correlate to the fingers of tetrapods.

However, the Rhipidistian fishes had many more fingers and finger bones. The loss and reduction of the numbers of bones of these animals is one of the way scientists can trace the more modern animals from the older ones. Modern living members of this group of extraordinary fishes include the coelacanths (considered to be living fossils) and the lungfishes. Lungfishes do not have gills, but have primitive **lungs** that enable them to remain out of water for extended periods of time. The mouth and skull structures of lungfishes are very similar to those of ancestral amphibians.

## Tetrapods in the Fossil Record

The fossil record of tetrapods is not complete. There are many gaps that prevent scientists from clearly understanding the relationship between ancestral amphibians and modern ones. In addition, the transitional form of animals representing the shift from amphibian to reptiles is still poorly understood. However, places like China, Europe, Mongolia, North America, and South America constantly provide new fossil information about the history of tetrapods, and tetrapod fossils have even been discovered in Greenland and Antarctica.

What is well-known about the history of tetrapods starts about 400 million years ago when the first terrestrial (no longer dependent on water for a complete life cycle) vertebrates appeared. By the beginning of the Triassic period many unusual amphibians ruled the land. A group of large and slow-moving creatures, the labyrinthodonts, and the smaller newtlike and salamander-like lepospondyls, dominated the swamps and humid environments of Earth.

Labyrinthodonts were named because the structure of their teeth whose outer and inner surfaces reminded researchers of a labyrinth, or maze. The teeth of lepospondyls were very simple and cone-shaped as compared to the labrinthodonts. The labyrinthodonts were some of the largest amphibians to have ever lived. Some genera, like the *Euryops*, were about 2 meters (six feet) long.

*Euryops* is considered one of many possible ancestors of modern amphibians The great diversity of these types of tetrapods has been vastly reduced. Only the salamanders and frogs are still living, but are found over

**body plan** the overall organization of an animals body

**cartilaginous** made of cartilage

**lungs** sac-like, spongy organs where gas exchange takes place

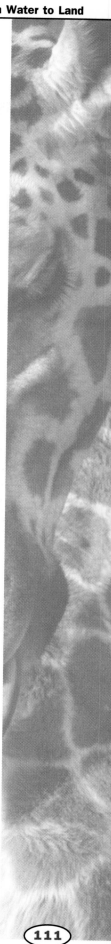

most of the world. Amphibians still have the primitive fishlike trait of laying eggs in water and have never lost their dependency on water-rich environments. As their skin does not retain moisture, they must live near a wet habitat to keep from drying out. They are good examples of the tetrapods' link to watery beginnings.

## On the Land and in the Sea

**amniotic** describes a vertebrate that has a fluid-filled sac that surrounds the embryo

The development of the **amniotic** egg and the growth of scales that prevented water loss allowed tetrapods to move into newer, more arid environments. An evolutionary explosion then occurred that produced the early ancestors of the turtles, crocodiles, lizards, snakes, dinosaurs, and even mammals. Many other tetrapods, like the pterosaurs, also emerged. In the marine environment, several tetrapods returned to the sea. Fierce marine reptiles like the mosasaurs and plesiosaurs found an abundant food source in the huge stocks of fishes in the oceans. These animals, although aquatic, were structurally tetrapods. They retained the tetrapod body plan of a thick, upper-arm bone connected to a girdle (hip and chest bones), two smaller bones, and a series of small bones for fingers. However, instead of appearing like an arm or a leg the bones were covered by tissue that formed a flipper.

Around 65 million years ago at the close of the Cretaceous period, a massive extinction killed many life forms including the large tetrapods (dinosaurs) and other animal and plant species. However, many life forms did not die out. Mammals, reptiles, and birds survived, and eventually became the most abundant tetrapods on Earth.

## Modern Tetrapods

The most familiar modern group of tetrapods is the mammals, which includes humans. These furred animals began their evolutionary history back with the beginnings of the dinosaurs. Early mammals were small, about the size of a rat. After the large Cretaceous extinction, mammals survived to become some of the largest tetrapods on Earth, including elephants and whales. Just as the marine reptiles had done, whales returned to the sea, developing fins as a secondary adaptation.

**habitat loss** the destruction of habitats through natural or artificial means

Tetrapods today can be found in nearly every environment. Unfortunately, as the world continues to change, the numbers of many types of tetrapods are declining. With increasing **habitat loss** and pollution many tetrapods are in danger of extinction. The loss of so many species of birds, reptiles, and mammals, such as the panda, is a serious threat to the continued survival of the tetrapods. SEE ALSO LIVING FOSSILS; PHYLOGENETIC RELATIONSHIPS OF MAJOR GROUPS.

*Brook Ellen Hall*

### Bibliography

Carroll, R. "Vertebrate Paleontology and Evolution." New York: W. H. Freeman and Company, 1988.

Lillegraven, J. A., Z. Kielan-Jaworowska, and W. Clemens. "Mesozoic Mammals; The First Two-Thirds of Mammal History." Berkeley: University of California Press, 1979.

Macdonald, D. "The Encyclopedia of Mammals," New York: Facts on File Publications, 1987.

# Threatened Species

Biodiversity—the vast numbers of species of plants and animals interacting on Earth—provides people with food, building materials, fibers, and medicines. Insects pollinate crops, wild plant **genes** reinvigorate domestic ones, forests provide breathable air, microorganisms create soil, and **aquatic** microorganisms cleanse and purify water. Without the millions of known and unknown species operating interdependently, life on Earth would be impossible for human beings. Yet life-forms are dying off at an unprecedented rate. Biologists estimate that there were 1 million fewer species in 2000 than in 1900.

The U.S. Endangered Species Act of 1973 defines an endangered species as one that is on the brink of extinction. A threatened species is defined as one with a good chance of becoming endangered in the foreseeable future. What pushes a healthy animal species into the threatened category? Almost all of the answers involve human beings. Either through misuse, thoughtlessness, or overconsumption of Earth's resources, humans are the source of the current mass extinction.

As human **populations** soar, more and more of the world's natural habitats are being destroyed. Forests are cleared for logging, shopping centers, and housing developments. Wetlands are drained for farms and factories. Valleys and rivers are flooded for hydroelectric power and recreation. Oil spills at sea threaten marine life. **Acid rain** from factories and automobile exhaust, **pesticides** scattered indiscriminately and chemical pollution of air and water further stress animal species. Legally protected animals are poached in wildlife preserves for their body parts, and ranchers and hunters exterminate grizzlies, wolves, large cats, and prairie dogs alike for trophies or because they consider them to be a nuisance.

One of the species headed for endangered status is the Pribilof seal. About 30,000 a year die entangled in plastic fishing lines and six-pack aluminum can holders. Most species require a critical number of members in order to successfully breed. The Key deer and Florida panther are diminishing below the breeding population (the number of individuals necessary for the species to breed successfully) because of deaths by automobile. Manatees are equally threatened due to collisions with motorboats.

Among the other threatened species are the whales, who are killed for their body parts. Mountain gorillas and pandas are threatened by shrinking habitat and food sources as human populations press in on their range. Asian tigers, South American jaguars, and snow leopards are disappearing as their pelts are sold as souvenirs. Sea turtles are hunted for their shells, reptiles for their skins, and rare birds are trapped to be sold as pets.

The numbers of unidentified species vanishing forever from the tropical rain forests, home to the greatest amount of life on Earth, has become a cause of great concern to environmentalists. The most disturbing phenomenon connected to the disappearing rain forests is that they are being destroyed to make room for ranches built to breed cattle that will eventually provide inexpensive meat products for developed nations.

**genes** segments of DNA located on chromosomes that direct protein production

**aquatic** living in water

**population** a group of individuals of one species that live in the same geographic area

**acid rain** rain that is more acidic than non-polluted rain

**pesticides** substances that control the spread of harmful or destructive organisms

The California Condor is a welcome instance of an animal almost ready to move from the federally protected Endangered Species list (placed there in 1967). Due to human intervention, the condor has successfully bred in captivity. Most importantly, birds reintroduced into the wild have been surviving.

**ecosystem** a self-sustaining collection of organisms and their environment

Like the dinosaurs 65 million years ago, humanity finds itself at the top of a food chain that is in the midst of a mass extinction. A mass extinction is a traumatic event. As the food chain falls apart, the survivors scramble to reassemble a workable **ecosystem**. For example, should the food of any common prey, such as rabbits or mice, become scarce, the numbers of those species will drop. This will set off a chain reaction that affects all the ground and air predators that feed on them, such as, coyotes, badgers, hawks and owls. The largest, most resourceful consumptive species do not survive. Ultimately, if humans continue to abuse the intricate ecological mechanisms that keep the world running smoothly, humans themselves may prove to be the most threatened species of all. SEE ALSO ENDANGERED SPECIES; EXTINCTION; HABITAT LOSS; HABITAT RESTORATION.

*Nancy Weaver*

**Bibliography**

Bailey, Jill, and Tony Seddon. *The Living World.* Garden City, NY: Doubleday, 1986.

Lampton, Christopher. *Endangered Species.* New York: Franklin Watts, 1988.

# Tool Use

Until quite recently, tool use was considered to be a uniquely human behavior. Early anthropologists taught that the use of tools was limited to *Homo sapiens* and used the presence of tools as an indicator of the presence of humans. Even after tool use by earlier Homo species was demonstrated, anthropologists still insisted that tool use made the species human.

More recently, scientists have observed many different animals using tools. Sea otters use rocks as anvils to break open shellfish. Galapagos finches mold twigs to probe holes in trees to obtain insect larvae. Egyptian vultures use rocks to crack open ostrich eggs. The burrowing wasp, *Ammophila*, uses a small pebble to hammer down the soil over its nest of eggs. And green herons use bait to attract small fish.

Since nonhuman animals were clearly observed to be using tools, anthropologists reconsidered their earlier position and decided that the *making* of tools was the uniquely human characteristic. However, in 1960 Jane Goodall and others observed chimpanzees in the wild breaking off twigs from trees, stripping away the leaves, and using the twigs to extract termites from their nests. Since the twig had to be modified by removing its leaves, this activity clearly demonstrated toolmaking. Captive Asian elephants have often been observed to use branches to swat flies. The branches are modified by shortening and removing side branches. Recently, wild Asian elephants have been seen demonstrating the same tool-using and toolmaking behavior.

**population** a group of individuals of one species that live in the same geographic area

The environment must possess two characteristics in order for tool use to evolve among a **population** of animals. First, there must be some evolutionary advantage in tool use. For example, the animal must be more successful at finding food, avoiding predators, or reproducing. Second, objects in the environment that will make useful tools must be available. Chimpanzees use twigs, wasps use small stones, and finches use cactus spines.

Without access to these objects, the tool-using behavior would never evolve. The following examples illustrate how Egyptian vultures, woodpecker finches, green herons, and chimpanzees gained an evolutionary advantage from tool use.

## Egyptian Vultures

An Egyptian vulture, *Neophron percnopterus*, has been observed breaking open ostrich eggs, too hard to open by pecking, by throwing a stone held in its beak at the egg shell. The bird's aim is quite good. According to reports by Jane Goodall, the vultures will wander as far as 50 meters (165 feet) from the egg to find a suitable rock.

The rock most often chosen is an egg shaped. This suggests that the vultures have modified an earlier behavior. Many birds throw or drop eggs to break them. John Alcock has suggested that the vultures originally threw or dropped eggs to break them open. The use of rocks to break the eggs open probably began when a vulture accidentally hit an egg with a rock. They then evolved from throwing the eggs to throwing rocks at the eggs. The movement the vulture makes when breaking an egg with a stone is almost the same as the movement the vulture makes when pecking at an egg to break it.

Observations by Chris Thouless and his co-workers of young, hand-reared vultures demonstrated that the stone-throwing behavior is innate or instinctive, not learned. Young vultures exhibit the behavior once it is linked to a food reward. In the wild, encountering a broken ostrich egg probably rewards young vultures. Thouless also observed that all vultures preferred to throw rounded-off, egg-like stones.

## Woodpecker Finches

Charles Darwin observed many species of finches inhabiting the Galapagos Islands and theorized that all of them evolved from a single-species population that had somehow made it to the island. The woodpecker finch, *Geospiza pallida*, that resides there is one of Darwin's finches. It is called the woodpecker finch because it has evolved the ability to pry insect larvae out of holes in a tree by using a cactus spine or other long, thin tool.

A woodpecker's long, barbed tongue enables it to extract grubs from branches without the assistance of a tool. The woodpecker finch has a short tongue, however, so it has developed the ability to grasp a cactus spine in its beak and pry grubs out of the hole. After extricating the grub, the finch holds the spine under its foot while eating the grub. The cactus spine is carried from branch to branch and used again and again.

Researchers have discovered that woodpecker finches are more likely to adopt tool-using behaviors as hunger increases. In an experiment a different finch, the cactus ground finch, was placed in a cage near a group of tool-using woodpecker finches. The large cactus ground finch does not normally use tools to probe for grubs in its natural environment but it apparently acquired the habit after observing the woodpecker finches at work. Other species of finches did not learn to use tools after observing woodpecker finches.

One researcher happened to observe a young woodpecker finch apparently perfecting the skill of using the cactus spine by observing another finch.

This woodpecker finch uses a twig to reach an insect.

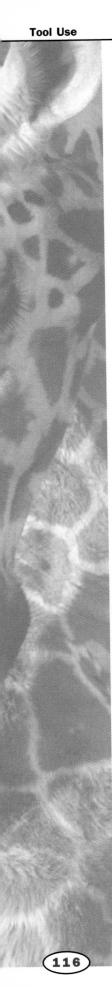

This young finch first tried to get grubs from a tree branch using its beak. Since this did not work, the young finch tried using a twig. Then it apparently observed another finch modifying a twig for use and copied the behavior. Scientists have also observed woodpecker finches modifying long cactus spines to form more manageable tools. This is a clear example of toolmaking.

## Green Heron

A few green herons have been observed dropping small objects onto the surface of the water. When small fish swim to the surface to investigate the object, the heron grabs the fish. This is apparently not **innate behavior** because only a few individual birds practice it. Attempts to teach herons to use bait have also been unsuccessful, so why a few do it is not clear.

It may be behavior learned through experience. Perhaps a bird accidentally drops a small object in the water and sees that small fish are attracted. If the heron is able to make the connection between dropping the bait and catching a fish, this indicates a level of cognitive ability beyond what is considered normal for these birds. Perhaps only exceptionally intelligent birds are able to learn the behavior.

## Chimpanzees

Chimpanzees have given us the first and clearest example of tool use and toolmaking in a nonhuman species. In Tanzania, chimps regularly construct tools from grass and twigs that they use to extract termites from termite holes. The chimp carefully selects an appropriate stem or twig, modifies it as necessary, and then uses its strong fingernail to dig a hole in the termite mound. As termites rush to repair the damage, the chimp carefully inserts the twig into the hole and just as carefully withdraws the twig. Invariably, several termites are clinging to the twig, which the chimp eats.

These insects are a good source of protein and fat, so they are a valuable addition to the chimp's diet. Wild chimpanzees have also been seen using sticks to get honey from beehives and dig up edible roots. They also use sticks to pry up the lids of boxes of bananas left by scientists.

Twig using is at least partially **learned behavior**. After extensive observations, Jane Goodall concluded that young chimps learn how to break twigs from trees, strip away the leaves, and insert them into termite holes by imitating adults. This is a complex and involved behavior. Without the adults to demonstrate, young chimps would probably never become skillful. However, the behavior is also at least partly innate. All young chimps play with sticks and twigs and entertain themselves by poking the sticks into holes.

If neither tool use nor toolmaking distinguish humans from other animals, is there any aspect of tool use that is still uniquely human? The answer is, possibly, no. It is conceivable that intelligence and tool use lie along a spectrum and humans just have more of the characteristic. The difference may be quantitative instead of qualitative. But, it is also possible that the regular, extensive use of tools to solve everyday problems is a distinctly human characteristic. The regular and consistent use of tools frees humans from limits imposed by our anatomy. It is even possible that regular tool

**innate behavior** behavior that develops without influence from the environment

**learned behavior** behavior that develops with influence from the environment

use has somehow encouraged the development of problem-solving skills.
SEE ALSO LEARNING.

*Elliot Richmond*

**Bibliography**

Alcock, John. *Animal Behavior: An Evolutionary Approach*, 6th ed. Sunderland, MA: Sinauer Associates, 1998.

Beck, Benjamin B. *Animal Tool Behavior: The Use and Manufacture of Tools by Animals*. New York: Garland, 1980.

Ciochon, Russell L., and John. G. Fleagle. *Primate Evolution and Human Origins*. Menlo Park, CA: Benjamin/Cummings, 1985.

Curtis, Helena, and N. Sue Barnes. *Biology*, 5th ed. New York: Worth Publishing, 1989.

Greenhood William, and Robert L. Norton. "Novel Feeding Technique of the Woodpecker Finch." *Journal of Field Ornithology* 70 (1999).

Goodall, Jane. (1986). *The Chimpanzees of Gombe: Patterns of Behavior*. Cambridge, MA: Belknap Press, 1986.

Goodall, Jane, and Hugo van Lawick. "Use of Tools by Egyptian Vultures, *Neophron percnopterus*." *Nature* (1966):468.

Lack, David. "Darwin's Finches." *Scientific American* (April 1953):68.

Roberts, William A. *Principles of Animal Cognition*. New York: McGraw-Hill, 1997.

Schultz, Adolph H. *The Life of Primates*. New York: Universe Books, 1972.

Thouless, C R., J. H. Fanshawe, and B. C. R. Bertram. "Egyptian Vultures *Neophron percnopterus* and Ostrich *Struthio-camelus* Eggs: The Origins Of Stone-Throwing Behavior." *Ibis* 131 (1989).

# Transport

Transport is the controlled movement of substances from one part of a cell to another, or from one side of a cell membrane to the other. Because each cell must maintain an internal environment different from the external environment, it must regulate the movement of ions, proteins, toxins, and other molecules both across the cell membrane and within its **cytoplasm**. This control over its molecular environment may be accomplished through a variety of measures, one of which is the establishment of a barrier membrane between the cell and the external world.

One such barrier, the **bilipid membrane** of a cell, is composed of two hydrophilic, or water-soluble, sheets of molecules separated by an intervening hydrophobic, also called oily or fatty, region. This property results from the structure of the **phospholipid** molecule composing the membrane: a polar, hydrophilic head region, and a nonpolar, hydrophobic tail region. The two layers of membrane are oriented so that the hydrophilic heads face the internal and external cell, and the fatty tails are positioned between the two head layers.

The structure of this membrane assures that any polar, water-soluble, molecules in the hydrophilic extracellular space will be unable to pass through the nonpolar, fatty, region within the membrane. If this barrier were completely impermeable, however, the cell would never be able to absorb nutrients or rid itself of wastes, let alone communicate with other cells. Thus the membrane is compromised by proteins that extend through both

**cytoplasm** fluid in eukaryotes that surround the nucleus and organelles

**bilipid membrane** a cell membrane that is made up of two layers of lipid or fat molecules

**phospholipid** molecules that make up double layer membranes; one end of the molecule attracts water while the other end repels water

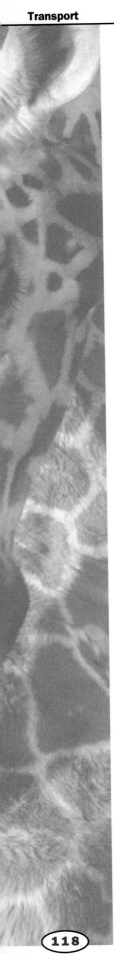

The functions of a bilipid membrane.

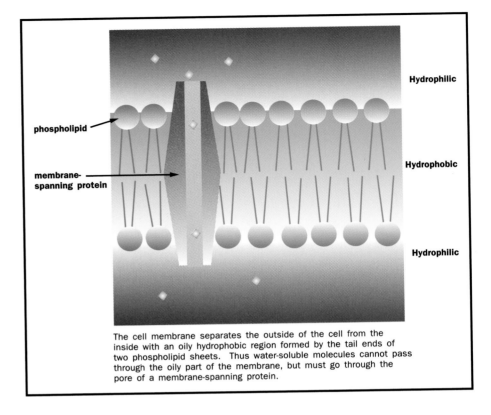

The cell membrane separates the outside of the cell from the inside with an oily hydrophobic region formed by the tail ends of two phospholipid sheets. Thus water-soluble molecules cannot pass through the oily part of the membrane, but must go through the pore of a membrane-spanning protein.

sides of the bilipid membrane. These proteins provide a watery pore region that connects the extracellular with the intracellular space, thereby allowing hydrophilic molecules to pass through.

There is a wide variety of membrane proteins, some of which are located within the outer cellular membrane, and some of which are imbedded in intracellular **organelles**. In addition to transport across membranes, substances must be transferred from one part of the cell to another, especially in very large cells and in single-celled organisms.

**organelles** membrane-bound structures found within a cell

## Osmosis

**osmosis** the diffusion of water across a membrane

The simplest kind of cellular transport is **osmosis**. Osmosis is the passage of water molecules through a semipermeable membrane from a wet environment (a region of high water concentration), to a dry environment (a region of low water concentration). The defining characteristic of a wet environment is a low concentration of dissolved solute in the water. A dry environment has a high concentration of solute. Cells store a high concentration of proteins and other molecules within the membrane. If a cell were removed from biological conditions and placed in distilled water (water containing no dissolved substances), the small water molecules would rush through the bilipid membrane into the relatively dry interior of the cell. The membrane would expand with the increased water intake until the cell exploded.

This scenario does not normally occur in nature for two reasons. First, organisms balance their internal osmolarity. Internal osmolarity is the ratio of dissolved substance concentration between the inside and the outside of the cell. This balance is accomplished by maintaining a concentration of

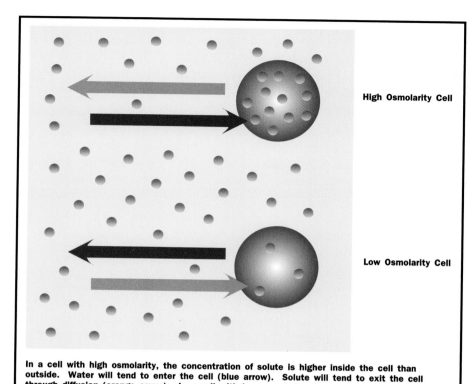

**High Osmolarity Cell**

**Low Osmolarity Cell**

In a cell with high osmolarity, the concentration of solute is higher inside the cell than outside. Water will tend to enter the cell (blue arrow). Solute will tend to exit the cell through diffusion (orange arrow). In a cell with low osmolarity, the concentration of solute is lower inside the cell than outside. Water will tend to exit the cell, and solute will enter through diffusion.

The osmosis process is the simplest kind of cellular transport.

many small ions and molecules in the extracellular space, the areas between cells within an organism. Second, cells store many of their proteins within vesicles, within membranes, and inside organelles; this decreases the apparent concentration of free-floating soluble molecules.

## Diffusion

The opposite of osmosis is diffusion, which means the passage of molecules from a region in which they are highly to a region in which they have a low concentration. This only occurs when the molecule has a concentration gradient, meaning that it exists in larger numbers per unit area in one location and in smaller numbers per unit area in an adjacent location. If a high concentration gradient is established, it means that the difference in concentration between two adjacent locations is great. In the case of a cellular membrane, this means that a certain substance is at very high concentrations on one side of the membrane and very low concentrations on the other side. The larger the concentration gradient, the stronger the driving force that powers the diffusion of molecules down the gradient. Whereas osmosis explains the movement of water molecules, diffusion explains the movement of other molecules within a liquid.

## Passive Transport

For passive transport, no additional energy is needed to transfer molecules across the membrane. Instead, the concentration gradient of that molecule provides a driving force from high to low concentration, pushing the

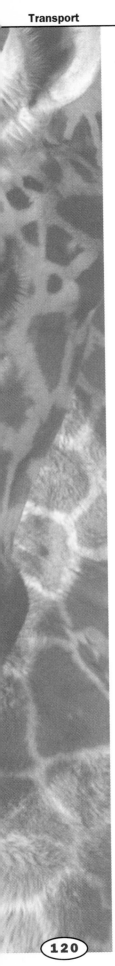

**facilitated diffusion** the spontaneous passing of molecules attached to a carrier protein across a membrane

molecule across the membrane. This process is also known as **facilitated diffusion** because the membrane protein facilitates the natural diffusion tendency of the molecule by providing safe passage across the hydrophobic region of the lipid bilayer. Many sugars and amino acids are transported from the gut into the cells lining the gut through this mechanism.

When the protein forming the pore is constantly open to diffusable molecules, it is called a channel. The size of the pore, and any hydrophilic regions within the pore, can selectively allow certain molecules to pass while barring others. A uniport protein must first bind a molecule and then undergo a conformational change (a change in the shape of the protein) before it releases the molecule on the opposite side of the membrane.

**smooth muscle** the muscles of internal organs that are not under conscious control

One example of a passive transport mechanism is the gap junction. This is a molecular conduit between two cells formed by a hexagonal array of rigid proteins that leaves a permanent pore open between two cells' cytoplasms. This opening allows inorganic ions and very small hydrophilic molecules to pass directly between cells. Gap junctions are particularly useful in heart muscle cells because they allow ions carrying electrical charge to flow directly from cell to cell, thereby inducing the **smooth muscle** fibers to contract in a coordinated motion. This simple method of communication makes it unnecessary for neuronal connections at each and every heart muscle cell to induce contraction at exactly the right time for a smooth muscle heartbeat. If gap junctions were blocked, heart muscle cells would contract at the incorrect time, or not at all, resulting in irregular or weak heartbeats, or even heart seizure, and death.

## Active Transport

**active transport** process requiring energy where materials are moved from an area of lower to an area of higher concentration

**adenosine triphoshate** an energy-storing molecule that releases energy when one of the phosphate bonds is broken; often referred to as ATP

Active transport requires that energy be expended to bring a molecule across the membrane. Most often **active transport** is used when the molecule is to be transferred against its concentration gradient. However, occasionally the target molecule is carried along its concentration gradient, when the gradient is not strong enough to ensure a sufficiently quick flow of molecules across the membrane. For uncoupled active transport, ATP (**adenosine triphosphate**) in the cell's cytoplasm binds to the carrier protein at the same time as the targeted molecule binds. ATP is a cellular molecule that contains a great deal of energy in its molecular bonds. The carrier protein breaks a phosphate off the ATP and uses the energy released from the broken bond to undergo a conformational change that carries the targeted molecule to the other side of the membrane.

**neuron** a nerve cell

One example of active transport is the sodium/potassium ATPase pump, by which sodium is transferred out of a **neuron** while potassium is transferred into a neuron, both counter to their individual concentration gradients. This is essential for maintaining electrically active nerve cells because it helps to establish a concentration gradient and electrical gradient across the cell membrane. When the cell is active, sodium and potassium are allowed to flow down their electrochemical gradients, whereas during periods of inactivity the sodium/potassium pump restores the resting state polarization of the cell.

There are two kinds of coupled active transport. Symport uses the concentration gradient of one molecule to help transport another molecule. Symport membrane proteins have two binding sites on the same side of the

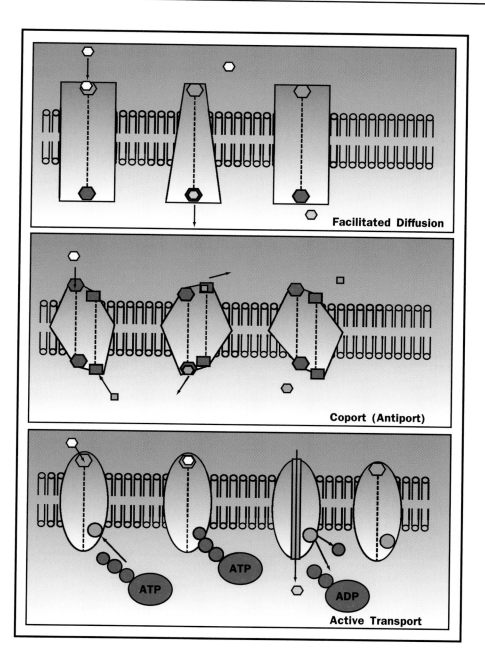

**Facilitated Diffusion**

**Coport (Antiport)**

**Active Transport**

The diffusion process describes the passage of molecules from a region where they are in high concentration to one in which they have a low concentration.

membrane. The targeted molecule binds at one site, and a coactivating molecule with a high concentration gradient favoring movement across the membrane binds at the other site. The driving force of the coactivating molecule causes a conformational change in the membrane protein, which transports both molecules across the membrane in the same direction. The other type of coupled transport is antiport, by which the coactivating molecule and the targeted molecule bind at sites on different sides of the membrane bilayer. The concentration gradient of the coactivating molecule still provides the energy for the conformational change, but in this case, the molecules are transported simultaneously in opposing directions across the membrane. Symport and antiport systems are also called carrier-assisted transport, because energy from the coactivator helps to carry the target molecule in an energetically unfavorable direction.

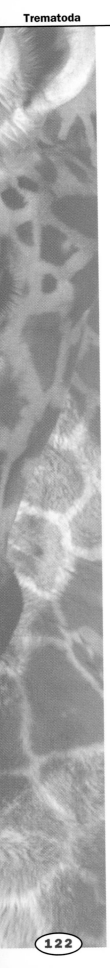

The most common method for cells to transfer very large molecules across the membrane is through the intermediary of a vesicle. Vesicles are small spheres of membrane that contain large molecules, toxins, nutrients, or signaling molecules. Proteins that are created in the endoplasmic reticulum can be packaged in vesicles in the golgi apparatus. Similarly, vesicles may contain transporter proteins similar to those located in the external cellular membrane, so that they can bring in certain cytoplasmic proteins. Vesicles called lysosomes process and sequester harmful metabolic side-products so that they do not do damage to organelles in the cytoplasm.

In each of these cases, vesicles are transported to the outer cell membrane, a process known as exocytosis. The vesicles travel along a system of structural proteins called microtubules with the help of an associated motor protein called kinesin. Kinesin molecules "grab" vesicles with their globular head region and then take turns binding to the microtubules, so that the net result is that of a cartwheeling vesicle moving in one direction. Some proteins can be transported by binding directly to microtubules and slowly "riding" them as they slide toward the membrane. When it arrives, the vesicle lipid bilayer fuses with that of the outer cell membrane, so that the internal contents of the vesicle are released into the extracellular space. In endocytosis, the mechanism is reversed. Pockets of cell membrane that have bound a particular molecule dimple inwards toward the cell cytoplasm forming a deep pit, and then pinch off so that a vesicle forms on the intracellular side. Phagocytosis refers to the consumption of small cells or cell-pieces by larger cells by surrounding and engulfing them.

*Rebecca M. Steinberg*

## Bibliography

Agutter, Paul S. *Between Nucleus and Cytoplasm.* London and New York: Chapman and Hall, 1991.

Weiss, Thomas Fischer. *Cellular Biophysics.* vol. 1: Transport. Cambridge, MA: MIT Press, 1996.

Wilson, G. *Exploitation of Membrane Receptors and Intracellular Transport Pathways.* New York: Elsevier Science Publishers, 1989.

# Trematoda

**acoelomate** an animal without a body cavity

**bilaterally symmetrical** an animal that can be separated into two identical mirror image halves

**dorsoventrally** flattened from the top and bottom

**vertebrates** animals with a backbone

**ventral** the belly surface of an animal with bilateral symmetry

Trematodes and flukes are the common names for the flatworms in the class Trematoda, phylum Platyhelminthes. Trematodes have most of the same features as other classes of Platyhelminthes. They are **acoelomate**, unsegmented, **bilaterally symmetrical** triploblasts that are flattened **dorsoventrally**. Trematodes do not have a respiratory system, but do have a mouth and primitive digestive and excretory functions, although these are not developed into full organ system as in higher **vertebrates** like mammals. They have a primitive nervous system with a ganglion, or brainlike structure, in the head region.

All trematodes are parasitic, and most are endoparasitic, living inside other animals. Trematodes infest various organs in a wide variety of animals. Adult trematodes have two specialized suckers. One is an oral sucker that surrounds the mouth. The other is a **ventral** sucker in the middle of the body that helps trematodes hang on to their host. Trematodes have simple sensory organs around the mouth, but do not have some of the more complex sensory organs found in other flatworms, such as the eye spots of turbellarians. The mouth of trematodes is a muscular pharynx, and the larvae and adult stages can suck their food from their host by grabbing on with their powerful mouths. Trematodes expel undigested material through their mouth because they do not have an anus. Nitrogenous waste excretion and other **osmoregulatory functions** are preformed by the protonephridium, which consists of flame cells and tubule networks, all of which act as a primitive kidney.

Most trematodes reproduce sexually. Most species are **hermaphroditic**, but a few have separate sexes. Trematodes have complex life cycles that include eggs, free-swimming ciliated larvae, and adults. Different life stages pass through one or more hosts.

The two major subclasses of Trematoda are best distinguished by the differences in their life cycles. Trematodes in the subclass Aspidobothria have only a single host in their life cycle. Trematodes in the subclass Digenea have a complex life cycle. These flatworms always pass through a mollusk in the first stage of their life cycle, have at least one additional host (but may have more), and complete their life cycle in the body of a vertebrate. Most trematodes are in Digenea, more than nine thousand known species.

Trematodes are some of the most harmful parasites of humans. Blood flukes (*Shistosoma*) infect more than two million people worldwide with a disease called as schistosomiasis. People become infected by working, bathing, or swimming in water containing **mollusks** such as snails that carry the flatworms. At the beginning of the trematodes' life cycle, fertilized flatworm eggs are passed with human feces into the water, where they infect the mollusk. Once the eggs find a mollusk host the eggs grow into larvae that infect humans when released from the mollusk. The larvae penetrate the skin of humans and migrate through the bloodstream to the liver where they mature into adult flatworms. The adult flatworms can migrate to other organs, and the accumulation of eggs in various organs causes symptoms including acute pain and diarrhea. The adult shistosomes ingest the red blood cells of their host, causing **anemia**. Schistosome larvae that infect birds can also cause problems for humans. Larvae free-floating in water are killed by an immune response when they enter human skin, but the decomposing larvae cause an infection that leads to the bumpy, itchy skin known as "swimmer's itch." Other trematodes can be a problem for humans because they infest domesticated vertebrates such as sheep and cows. SEE ALSO PHYLOGENETIC RELATIONSHIPS OF MAJOR GROUPS.

*Laura A. Higgins*

A Bilharzia blood fluke during schistosomiasis.

**osmoregulatory functions** controlling the water balance within an animal

**hermaphroditic** having both male and female sex organs

**mollusks** a large phylum of invertebrates that have soft, unsegmented bodies and usually have a hard shell and a muscular foot; examples are clams, oysters, mussels, and octopuses

**anemia** a condition that results from a decreased number of red blood cells

### Bibliography

Anderson, D. T. ed. *Invertebrate Zoology*. Oxford: Oxford University Press, 1998.

Barnes, Robert D. *Invertebrate Zoology*, 5th ed. New York: Saunders College Publishing, 1987.

Campbell, Neil A., Jane B. Reece, and Lawrence G. Mitchell. *Biology*, 5th ed. Menlo Park, CA: Addison Wesley Longman, Inc., 1999.

Purves, William K., Gordon H. Orians, H. Craig Heller, and David Sadava. *Life: The Science of Biology*, 5th ed. Sunderland, MA: Sinauer Associates Inc. Publishers, 1998.

# Triassic

The Triassic period is the first of the three divisions of the Mesozoic era, which is known as "The Age of Reptiles." The period lasted for 37 million years, from 245 to 208 million years ago. The Triassic is named after a tricolor sequence of red, white and brown rock layers found in Germany.

Towards the end of the Paleozoic era, which preceded the Triassic, Earth's great crustal plates collided to form the supercontinent Pangaea. Early in the Triassic, Pangaea began to break apart again in a process that is still going on. North America, Europe, and Asia split away as one continent (called Laurasia) from South America, Australia, India, Africa, and Antarctica (known as Gondwanaland). These giant continents continued to break apart into the land masses we have today.

The Triassic began with a relatively warm and wet **climate**. However, deposits of fossilized sand dunes and **evaporites** (rocks formed from evaporation of salty and mineral-rich liquid) found in later Triassic strata suggest that the general climate was hot and dry, although some areas may have had defined rainy seasons.

Fossils found in Triassic rocks suggest that this was a period of transition, in which older forms of plants and animals died out and new ones began to appear. A major extinction had occurred at the end of the Paleozoic. Over 90 percent of marine invertebrate species (animals without a backbone) became extinct, as well as many other species of land plants and animals. Scientists hypothesize that a combination of volcanic eruptions, a global drop in sea level, climate change, or loss of habitat during the formation of Pangaea may have contributed to the extinctions.

The Triassic marked the beginning of important advances in plant life. The conifers (including pine trees) appeared to join the already flourishing ferns, cycads (a palmlike plant), horsetail rushes, and now-extinct species like seed ferns. Triassic plants had thick waxy coverings and did not usually grow as tall as modern trees. Scientists hypothesize that the tough coverings on these plants (perhaps developed to keep from drying out in the warm climate) contributed to the development of larger, blunt, rounded teeth (designed for tearing, shredding and chewing) in many animals, including dinosaurs.

Early Triassic oceans contained many different **invertebrates** than before. When seas reflooded the continents during the early Triassic period many changes took place in the composition of species. Many species died out following the Permian extinction and different species began to fill ecological niches. Modern reef-forming corals replaced earlier, more primitive forms.

Clams, snails, scallops, and other **mollusks** weathered the Permian extinction and replaced **brachiopods** as the most common shelled marine invertebrates. Two groups of bryozoans (small seaweed-like colonies attached to objects in shallow seawater), cryptostomate and fenestarte, became extinct. Many families of brachiopods (animals with two shells situated on the top and bottom of the animal) including *productaceans, chonetaceans, spiriferaceans,* and *richthofeniaceans* became extinct. The entire group of trilobites (early **arthropods** with three lobes—head, abdomen and tail that burrowed

**climate** long-term weather patterns for a particular region

**evaporites** rocks formed from evaporation of salty and mineral-rich liquid

**invertebrates** animals without a backbone

**mollusks** a large phylum of invertebrates that have soft, unsegmented bodies and usually have a hard shell and a muscular foot; examples are clams, oysters, mussels, and octopuses

**brachiopods** a phylum of marine bivalve mollusks

**arthropods** members of the phylum of invertebrates characterized by segmented bodies and jointed appendages such as antennae and legs

in mud or sand) died out during the Permian extinction. Calms, snails, scallops, and other modern mollusks weathered the Permian extinction. Sea stars, urchins, and sand dollars became the dominant **echinoderms**, over **crinoids** and blastoids.

Crinoids and blastoids were two groups that been previously very successful during the flower-like animals known as stalked echinoderms who attached their stalks (or bodies) to the sea floor. They had multiple food gathering "arms" that allowed them to filter water. Fishes continued to evolve and became better adapted to their watery environments. The **cartilaginous** fishes, represented by sharks and rays, had skeletons of **cartilage**, with bony teeth and spines.

The skeletons of bony fishes were primarily of bone instead of cartilage. An interesting group of bony fishes included the lungfish and the lobefin fishes. These ancestors of the amphibians had a backbone and small limbs for support, along with an air bladder (a primitive lung) that enabled them to breathe out of water for short periods.

Although, not as numerous as during the Permian period, the amphibians continued to be well represented and diverse.were diverse. One group of common Triassic amphibians was the labyrinthodonts ("labyrinth teeth"). These flat-headed creatures grew several feet in length, had sharp, conical teeth with deeply folded enamel (hence the name of the group), small limbs and very weak backbones, and spent their time in the swampy backwaters

This museum exhibit depicts two Triassic period dinosaurs, Plateosaurus and Yaleosaurus.

**echinoderms** sea animals with radial symmetry such as starfish, sea urchins, and sea cucumbers

**crinoids** echinoderms with radial symmetry that resembles a flower

**cartilaginous** made of cartilage

**cartilage** a flexible connective tissue

of Triassic rivers. The labyrinthodonts became extinct during the Triassic, but other amphibians, including frogs, became established.

One landmark of animal evolution during the Triassic was the success of the reptiles. Unlike amphibians, which must stay near water for much of their lives, reptiles were completely adapted to life on land and so occupied a variety of habitats, ranging from semidesert to dry uplands, marshes, swamps, and even oceans. Ocean reptiles included the dolphin-like ichthyosaurs, turtles, and large-bodied, long-necked, paddle-flippered reptiles known as plesiosaurs. Except for the turtles, all these marine reptiles are now extinct.

**vertebrates** animals with a backbone

Triassic land **vertebrates** were dominated by two groups. In the early Triassic, the synapsids, the group that includes mammals and their close relatives, were the most common. One group of synapsids, the "mammal-like reptiles," shared the characteristics of both mammals and reptiles and ultimately gave rise to mammals. True mammals—small, shrewlike creatures—did not appear until the late Triassic. The other large group of terrestrial vertebrates during the Triassic was the archosaurs, which includes dinosaurs, crocodiles, and pterosaurs (flying reptiles), as well as several other now-extinct forms.

**bipedal** walking on two legs

**carnivorous** term describing animals that eat other animals

The first dinosaurs appeared in the late Triassic. Early dinosaurs include *Eoraptor*, *Herrerasaurus*, *Staurikosaurus*, and *Coelophysis*. These creatures were small, **bipedal**, and **carnivorous**. By the end of the Triassic, dinosaurs had become the most common land vertebrates, along with the pterosaurs, crocodiles, and crocodile-like creatures known as phytosaurs.

*Leslie Hutchinson*

**Bibliography**

Lane, Gary A., and William Ausich. *Life of the Past*, 4th ed. Upper Saddle River, NJ: Prentice Hall, 1999.

# Trophic Level

**zooplankton** small animals that float or move weakly through the water

**autotroph** an organism that makes its own food

**ecosystem** a self-sustaining collection of organisms and their environment

**herbivore** animal who eats plants only

**chemotrophs** animals that make energy and produce food by breaking down inorganic molecules

A trophic level consists of organisms that get their energy from a similar source. Each step in a food chain is a trophic level. A food chain is a series of organisms each eating or decomposing the preceding organism in the chain. For example, in a lake phytoplankton are eaten by **zooplankton** and zooplankton are eaten by small fish. A food web is similar to a food chain, but in a food web there are many interconnected and interacting food chains. In a typical food chain, a producer or **autotroph** is the source of solar energy that powers the **ecosystem**. For example, in a grazing food web a **herbivore** eats living plant tissue (the producer) and is eaten in turn by an array of carnivores and omnivores. In contrast, a detrivore harvests energy from dead organic material and provides energy for a separate food chain.

There are communities of organisms surrounding deep-ocean hydrothermal vents that obtain their energy from bacteria (known as **chemotrophs** or **chemosynthetic autotrophs**) that harvest heat energy and store it in chemical bonds. These communities are the rare exception.

A simple food web.

**Photosynthesis** is the ultimate source of energy for every other ecosystem on our planet. **Producers** (autotrophs, or **photosynthesizing autotrophs**) use photosynthesis to harvest energy from the sun. All other organisms obtain their energy, directly or indirectly, from the autotrophs. All of the other organisms in the food chain are **consumers** (known as heterotrophs). The primary consumers eat the producers. Secondary consumers eat the primary consumers, and so on. For example, in a grassland ecosystem, grass is the producer. Grasshoppers are primary consumers. Shrews that prey on the grasshoppers are secondary consumers. Owls, hawks, and snakes prey on the shrews, so they are tertiary consumers. Of course, hawks also prey on snakes and grasshoppers, so the connections get complicated and are usually described as a web of relationships or a food web.

A food chain involves a transfer of matter and energy from organism to organism. As energy is transferred through the food chain or food web, some energy is converted to waste heat at each transfer. The quantity of energy lost is so great that food chains rarely involve more than four or five steps from consumer to top predator. Each level in a food chain or food web is known as a trophic level, a group of organisms that all consume the same general types of food in a food web or a food chain. In a typical food web, all producers belong to the first trophic level and all herbivores (primary consumers) belong to the second trophic level. Using the same grassland as an example, the second trophic level would be all of the herbivores that eat the grass. This group can include a wide variety of different organisms. In the original grasslands of the central United States, the second trophic level included grasshoppers, rabbits, voles and other small rodents, prairie dogs, and American bison. Since they all eat the same grass, they are all at the same trophic level, despite their differences in size, reproductive habits, or any other factors.

The second trophic level in an ecosystem is relatively easy to identify, because the organisms in this level all obtain their energy directly from the autotrophs at the first level. After the second level in a food web, the situation becomes progressively more complex. Many organisms obtain energy from several different sources at different trophic levels. For example, foxes are opportunistic omnivores. They will eat fruits, small herbivores, and small carnivores. Likewise, many birds eat seeds and fruits in one season and switch to eating insects in a different season.

**chemosynthetic autotrophs** organisms that uses carbon dioxide as a carbon source but obtain energy by oxidizing inorganic substances

**photosynthesis** converting sunlight to food

**producers** organisms which make up the level of an ecosystem that all other organisms ultimately depend on; usually these are plants

**photosynthesizing autotrophs** animals that produce their own food by converting sunlight to food

**consumers** animals that do not make their own food but instead eat other organisms

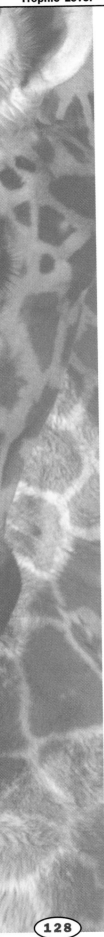

Ecological pyramids illustrate the amount of energy available at each of the four levels of an ecosystem. Redrawn from Johnson.

**biomass** the dry weight of organic matter comprising a group of organisms in a particular habitat

Another problem in classifying trophic level arises because the energy available at a given trophic level includes many different forms. Plant tissue includes wood, nectar, pollen, seeds, leaves, and fruit. No animal eats all of these different forms of plant tissue. Animals with very different ecological characteristics exploit these various tissues. Termites eat wood; fruit bats eat fruit. Termites offer an additional complication, because it is the organisms in the termite's gut that digest the cellulose who are the actual second-level consumers.

How energy flows through an ecosystem depends on the nature of the producers at the first trophic level. These producers support the entire ecosystem, so their abundance and energy content per kilogram determine the overall energy flow and **biomass** of the ecosystem. Organisms that live on land expend much of their energy in building supporting structures. These supporting structures are not generally available as an energy source to consumers. For example, in a forest, both matter and energy accumulate in the form of wood fibers that cannot be eaten by most animals. On the other hand, grasses have little supporting structure. The herbivores that consume grass are able to eat all of the above ground parts of the plant. With the aid of the specialized bacteria in their guts, grassland herbivores are able to harvest more energy per kilogram of plant material present.

A grassland ecosystem—or any ecosystem—can be represented by an energy pyramid. The base of the pyramid is the community of producers, including various species of grass. The primary consumers, grasshoppers, rodents, rabbits, and bison make up the second level of the pyramid—herbivores. The third level of the pyramid is the secondary consumers, predators that prey on the herbivores. The producers far outnumber the herbivores who far outnumber the carnivores, so the grassland pyramid has a broad base and a narrow tip.

**aquatic** living in water

**population** a group of individuals of one species that live in the same geographic area

In **aquatic** environments away from the shoreline, the situation is reversed. The primary photosynthesizing organisms are algae. Small herbivores (grazers) consume the entire organism and harvest almost all of the energy. These grazers decimate the algae **population**, keeping it relatively small. In this ecosystem, most of the energy (and matter) is stored in the second trophic level—the grazers. However, the high reproductive rate and short life cycle of the algae keep the population at a level sufficient to support the grazers. SEE ALSO BIOMASS; FOOD WEB.

*Elliot Richmond*

### Bibliography

Curtis, Helena, and N. Sue Barnes. *Biology*, 5th ed. New York: Worth Publishing, 1989.

Miller, G. Tyler, Jr. *Living in the Environment*, 6th ed. Belmont, CA: Wadsworth Publishing, 1990.

Johnson, George B. *Biology: Visualizing Life*. New York: Holt, Rinehart and Winston, Inc. 1998.

# Turbellaria

The class Turbellaria is the most primitive group within the phylum Platyhelminthes, the flatworms. Turbellarians share some important characteristics with other Platyhelminthes. All flatworms are flattened **dorsoventrally**. They are **bilaterally symmetrical**, are unsegmented, and are **acoelomates**, which means they do not have a body cavity. Turbellarians are solid because all the space around their digestive cavity is filled with muscle and other tissue. Turbellarians do not have a respiratory or circulatory system, they exchange gases by diffusion through all their cells. They have a muscular mouth, called a pharynx, as well as a saclike digestive cavity. Turbellarians also have an **osmoregulatory system** called the protonephridium. This system is made up of tubules, a network of little tubes, and specialized cells called flame cells.

The turbellarian nervous system includes a primitive brainlike structure in the head region, called a ganglion. This ganglion is formed by the thickening of the **anterior** part of the **ventral** nerve cords. The head region also has specialized sensory organs, which are more complex in turbellarians than in other flatworms. These organs include eye spots, which are composed of photoreceptors that detect light and are **tactile** and chemical sensory organs that help turbellarians find food. Movement is assisted by receptors that help maintain balance as well as detect movement. The sensory organs of turbellarians are more complex than those of other flatworms because turbellarians are free-living; all other flatworms are parasitic. The free-living turbellarians are ancestors of the parasitic flatworms; parasitism evolved as a specialized form of feeding and reproducing from the scavenger lifestyle of turbellarians. Turbellarians eat both living and dead animal material. Some turbellarians secrete digestive **enzymes** onto their food, then ingest the already-digested food particles through their pharynx. Others digest food in their digestive cavity. All flatworms must expel undigested food out of their mouth; they do not have an anus.

Turbellarians move around using **cilia** on their epidermis or by undulating their body with their muscles. Most turbellarians live in water, either fresh or salt water. A few species live on land in damp habitats like leaf litter. Turbellarians reproduce by **fission** and regeneration, or sexually. Turbellarians that reproduce sexually are hermaphroditic—sperm from one animal will fertilize eggs from another, and the eggs then hatch into small

**dorsoventrally** flattened from the top and bottom

**bilaterally symetrical** an animal that can be separated into two identical mirror image halves

**acoelomates** animals without a body cavity

**osmoregulatory system** a system that regulates the water balance between an organism and its environment

**anterior** referring to the head end of an organism

**tactile** the sense of touch

**enzymes** proteins that act as catalysts to start biochemical reactions

**cilia** hair-like projections used for moving

**fission** dividing into two parts

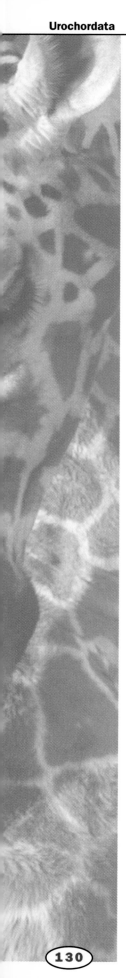

turbellarians. When reproducing by fission and regeneration, the tail end of the individual turbellarian adheres to a substrate and the head region pulls away from the tail. This eventually splits the flatworm in two, and each piece regenerates the end that is missing. Turbellarians are the only flatworms that can reproduce by fission.

There are more than 4,500 species of turbellarians. Most are less than 5 millimeters (0.2 inches) long, and many are microscopic in size. Planarians (*Dugesia*) are largest turbellarians; they can grow up to 0.5 meter (20 inches) long. SEE ALSO PHYLOGENETIC RELATIONSHIPS OF MAJOR GROUPS.

*Laura A. Higgins*

### Bibliography

Anderson, D. T., ed. *Invertebrate Zoology*. Oxford: Oxford University Press, 1998.

Barnes, Robert D. *Invertebrate Zoology*, 5th ed. New York: Saunders College Publishing, 1987.

Campbell, Neil A., Jane B. Reece, and Lawrence G. Mitchell. *Biology*, 5th ed. Menlo Park, CA: Addison Wesley Longman, Inc., 1999.

Purves, William K., Gordon H. Orians, H. Craig Heller, and David Sadava. *Life: The Science of Biology*, 5th ed. Sunderland, MA: Sinauer Associates Inc. Publishers, 1998.

**notochord** a rod of cartilage that runs down the back of Chordates

**dorsal** the back surface of an animal with bilateral symmetry

**metamorphose** drastically changing from a larva to an adult

**sexual reproduction** a reproduction method where two parents give rise to an offspring with a different genetic makeup from either parent

**metamorphosis** a drastic change from a larva to an adult

# Urochordata

Urochordates are small marine animals with larvae that swim freely and adults that attach themselves to the ocean floor. The 1,300 species of urochordates, like all members of the phylum Chordata, possess four characteristic anatomical structures as embryos: a flexible body-length rod (the **notochord**) that provides resistance against muscular contractions and allows for more efficient movement; a **dorsal**, hollow, nerve cord that forms the central nervous system; slits in the beginning of the digestive tract (the pharynx) that allow filter feeding and gas exchange; and a postanal tail. Urochordates, commonly known as tunicates, differ from other chordate subphyla (Cephalochordata and Vertebrata) in that the adult form has no notochord, nerve cord, or tail. In fact, an adult tunicate is an immobile, filter-feeding marine animal that in some ways looks more like a sponge or a mollusk than a chordate.

Tunicate larvae look much more like other chordates than adult tunicates do. Larvae swim using their notochord as structural support for strong, wavelike body movements. However, a larval tunicate cannot eat because both ends of its digestive tract are covered by a skinlike tissue called the tunic. The sole purpose of the larva is to find a place to attach its tail to the ocean floor, where it can **metamorphose** into an adult and begin **sexual reproduction**. **Metamorphosis** involves the loss of the notochord, nerve cord, and tail, and a twisting of the body so that the mouth and the anus both point away from the attachment. The adult can then start filter feeding. It does this by sucking water into its mouth through a pharynx with slits that captures food and spitting the water back out a different opening called the atriopore.

Tunicates might seem like an evolutionary off-shoot, given that the adult form is so different from other chordates. However, based on the similarities between larval tunicates and embryonic **vertebrates**, evolutionary biologists suspect that the first vertebrate chordates were very much like modern-day tunicates. They propose an evolutionary mechanism called paedomorphosis, in which the larval form evolved the ability to reproduce before metamorphosis. No longer requiring a suitable ocean floor to reproduce, these protovertebrates would have been free to exploit and adapt to new niches, eventually giving rise to the vertebrate skeleton.

To say that our vertebrate ancestors resembled tunicates is not to say that those ancestors were tunicates. Modern-day tunicates have had 500 million years to evolve since the days of the ancestors they share with vertebrates. Conceivably, then, protovertebrates did not have an adult form like that of the tunicate, and this adult form evolved later. Without fossil evidence of the earliest vertebrates, it may never be known whether or not they resembled tunicate adults. SEE ALSO PHYLOGENETIC RELATIONSHIPS OF MAJOR GROUPS.

*Brian R. West*

### Bibliography

Campbell, Neil A. *Biology*, 2nd ed. Redwood City, CA: Benjamin/Cummings Publishing Company, Inc., 1990.

Curtis, Helena, and N. Sue Barnes. *Biology*, 5th ed. New York: Worth Publishers, 1989.

Ridley, Mark. *Evolution*, 2nd ed. Cambridge, MA: Blackwell Science, Inc., 1996.

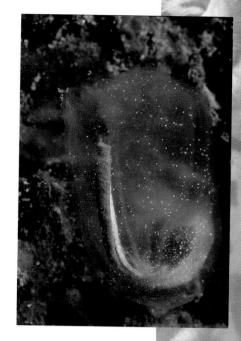

A transparent sea squirt.

**vertebrates** animals with a backbone

# Vertebrata

The vertebrates are commonly called "animals with backbones," but this is a simplified description of a group of animals who are the most anatomically and functionally diverse on Earth. As with most major groups of animals, their beginnings are not known. However, scientists have constructed a theory of the origins of vertebrates that is generally well accepted by the academic community.

Scientists base their model of a hypothetical vertebrate ancestor on several primitive living vertebrates. At the forefront is *Amphioxus*, whose body shape is close to what scientists believe resembles that of the ancestral vertebrate. Many researchers have successfully examined the body of this small marine animal and it has become a popular organism for study in biology classes.

Based on *Amphioxus*, scientists believe that the first vertebrates had a fishlike body with individual segmented muscles along its entire length. They were small ocean dwellers who lived close to the bottom and used their muscles to contract and move their bodies and tail in a side-to-side motion. This action propelled them through the water. They had no distinct head. The brain of early vertebrates is a somewhat controversial topic among scientists, but most researchers agree that the early forms had a brain that was more complex than the simple brain of *Amphioxus*. However, the brain was not the highly complicated structure we see in most vertebrates today. The early brain probably carried out only the most basic of body and sensory functions. As with many animals with cerebral

Like other vertebrates, this Cactus Finch has a bilaterally symmetrical body. If divided down the middle, the bird's body will split into two equal halves.

**lateral line** a row of pressure sensitive sensory cells in a line on both sides of a fish

**cartilage** a flexible connective tissue

**cartilaginous** made of cartilage

head ganglia, nerve tracts emerged from the brain and ran along the length of the body. Scientists assume the nerve tracts responded to sensory stimuli. The **lateral line** system, present in most fish, was most likely present in early vertebrates.

Just as in *Amphioxus*, a dorsal (top of body) hollow nerve cord ran the entire length of the body. This nerve cord was supported by an important evolutionary structure made of **cartilage** called the notocord. The notocord, and its role in the evolution of vertebrates, is one of the most important characteristics distinguishing vertebrates from nonvertebrates. Although there are many other characteristics that help to classify the group, the notocord is the structure from which the backbone is believed to have evolved and is the structure from which vertebrates get their name.

As vertebrates became more specialized and increased their ability to move and sense their environment, the brain and spine became more complex. In looking at the growth of fetal vertebrates it has been shown that the developing muscles place a strain on the **cartilaginous** notocord. As the strain on the notocord increases with growth of the muscles, deposits of bone replace the cartilage, giving the rod greater strength. This process of replacement eventually produces the bones known as the vertebrae. Each vertebra in a primitive vertebrate corresponds to an individual set of muscles. This pattern is harder to recognize in more derived vertebrates, like mammals and birds, but it is there nonetheless.

In primitive vertebrates, the mouth is a simple oral opening that leads to the **gill slits** and digestive system. As the vertebrates continued to evolve, the oral cavity was replaced by a more specialized mouth and gill apparatus. Although sharks and their relatives have a primitive type mouth and gills, bony fishes such as salmon and perch have developed complicated gills with a bony covering called an operculum. Sharks are primitive vertebrates in that they do not develop bony skeletons, but even so, the cartilage structure of vertebrates is easy to see.

As vertebrates become complicated in body structure, the mouth becomes a very characteristic structure. Many have teeth that are actually modified body scales. The increasing specialization of the teeth, such as the pointed, socket-bound teeth of the reptile or the many cusped teeth of the mammals, is a major trait on which groups of vertebrates are identified. Birds have no teeth whatsoever.

Early vertebrates had a simple mouth opening through which they gulped food like a frog or fish. This structure was not only poorly adapted for capturing and holding on to active prey but also prevented the animal from breathing while trying to feed. Hunting and swallowing quick prey, like some flying or hopping insects, was problematic to the ill-equipped vertebrates. As a response to increasingly swift food sources that were adapting to life on land, the vertebrates became swifter and more dangerous. The increased specialization of the mouth proved to be an advantage for the group in capturing food.

As they struggled with larger or stronger prey, it was necessary to bite and hold on to wriggling and unwieldy insects. It was hard for them to breathe and many primitive vertebrates were unable to capture these more agile animals.

A major evolutionary trend in the vertebrates was the development of the secondary palate in the mouth, a platform of bone that separates the nasal cavity from the mouth. Mammals, including humans, have a secondary palate that allows for breathing while feeding. This means that the hunter can bite and hold onto its prey and still breathe. It can chew or tear at its food instead of gulping like a crocodile. Lions are an excellent example of how the secondary palate helps the lion to bite and hold onto its intended victim until it is dead and then tear off portions for eating.

It is difficult to provide a generalized summary of the characters of all vertebrates. The group is extremely diverse and includes fish, sharks and rays, amphibians, reptiles, birds, and mammals. However, there are several characteristics that are common to all vertebrates and four that are completely exclusive to the group.

All vertebrates are bilaterally symmetrical—they have two sides which are identical to each other in one plane only. A vertebrate can be divided down the middle, or **sagital plane**, to produce two equal halves. However, if it is divided down the side, or **transverse plane**, the sides will not be identical.

The notocord, or skeletal rod, and the **dorsal** hollow nerve cord are present in all vertebrates. These two characters are unique to vertebrates.

**gills** site of gas exchange between the blood of aquatic animals such as fish an the water

**sagital plane** plane long ways through the body

**transverse plane** a plane perpendicular to the body

**dorsal** the back surface of an animal with bilateral symmetry

**pharyngeal** having to do with the tube that connects the stomach and the esophagus

**prehensile** adapted for siezing, grasping, or holding on

**ventral** the belly surface of an animal with bilateral symmetry

**exoskeleton** hard outer protective covering common in invertebrates such as insects

Other characteristics unique to the vertebrates are **pharyngeal** (at the sides of the pharynx, or throat) gill slits and a tail behind the anal opening. The presence of the tail may seem an obvious trait, but no other group of animals has a structure that can be identified as an actual tail behind the anal opening. Some insects have bodies that extend beyond the anus, but they do not have tails. Vertebrate tails can move and provide locomotion, or balance and support, as in birds and dinosaurs. In many vertebrate groups, such as monkeys, which use their **prehensile** tails for swinging through trees, the tail can act as an extra appendage.

The vertebrate circulatory system is always closed, but this is not unique to the group. However, the vertebrate heart is always located in a **ventral** position and the digestive system is complete. Surprising to many, certain vertebrates, especially extinct forms, have an **exoskeleton**. Early fishes, called ostracoderms and placoderms, had bony exoskeletons that protected their head and sensory areas.

The vertebrates have become a highly successful group of animals with an interesting and exciting evolutionary story. Because humans are vertebrates, they have a natural and continuing curiosity about their predecessors who, most likely, had their beginnings about 500 million years ago in the seas of Earth.

*Brook Ellen Hall*

# Veterinarian

Veterinarians are doctors who diagnose and treat the diseases and injuries of animals. Their duties vary depending on the specific type of practice or specialization. In North America, 80 percent of veterinarians are in private practice. The remaining 20 percent work at zoos, inspect meat and poultry for the federal or state government, teach at veterinary universities, or conduct research. When serving in the U.S. Army, Air Force, or Public Health Service, veterinarians are commissioned officers.

Veterinarians who practice general veterinary medicine are similar to family practitioners. They carry out a number of health services, including giving general checkups, administering tests and immunizations, and advising pet owners on feeding, exercising, and grooming. Veterinarians are called upon to perform routine spaying or neutering, as well as surgery to correct a health problem. Most veterinarians treat small animals and pets exclusively, but others specialize in larger animals such as cattle, horses, and sheep. Veterinarians usually have set office hours, although they may be called to take care of an emergency at any time. Large-animal veterinarians may also be required to visit a farm or ranch when an animal is injured, ill, or expected to give birth under difficult circumstances.

A veterinarian must get along with animals and should have an interest in science. There are a limited number of colleges of veterinary medicine in the United States, and entrance is competitive. Individuals must complete six years of college to receive a Doctor of Veterinary Medicine (D.V.M.) degree. Coursework during the first two years emphasizes physical and biological science. The remaining four years involve classroom work; practical experience in diagnosing and treating animal diseases; surgery; lab

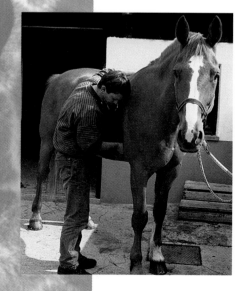

A veterinarian listens to this horse's heartbeat during a check-up.

courses in anatomy and biochemistry; and other scientific and medical studies. In addition, veterinarians are required to pass an examination in the state, including the District of Columbia, in which they wish to practice.

*Stephanie A. Lanoue*

### Bibliography

*Encyclopedia of Careers and Vocational Guidance.* Chicago: Ferguson Publishing Company, 2000.

Field, Shelley. *100 Best Careers for the Year 2000.* New York: Prentice Hall, 1992.

*VGM's Careers Encyclopedia,* 4th ed. Chicago: NTC Publishing Group, 1997.

# Viruses

Viruses are infectious agents that have no **organelles** or reproductive machinery of their own. Viruses cannot duplicate their DNA or RNA, nor can they translate their genetic information into protein. Essentially, they are small bags of **genes** that typically encode a comparatively small number of proteins. For example, the human immunodeficiency virus (HIV) is composed of only nine genes, yet with these simple nine bits of protein it can wreak havoc on the human immune system. Others, such as herpes simplex or adenovirus, can have large **genomes** with dozens of genes. Simple or complex, though, all viruses have the same function. As they cannot make protein or reproduce on their own, viruses must force bacteria or animal cells to do their work for them. A virus is, simply put, a genetic parasite.

As it does not have to sustain other energetically expensive cellular processes, a virus has a very simple structure. It is usually composed of nucleic acids (DNA or RNA) and a protein coat. This coat may be a very primitive covering (called a capsid) or it may be a complex structure derived from a host's membrane (called an envelope). Viral envelopes may possess a variety of receptors and decoys designed to fool the host's immune system. Avoiding detection is one of a virus's main tasks, and viruses may rely on dummy surface molecules or manipulating the immune system to do so. For example, HIV convinces infected cells to stop producing molecular flags that indicate infection to the rest of the body. In order to be successful, a virus must defeat the immune system and reproduce itself efficiently.

A viral life cycle generally has five distinct phases:

1. Attachment: The virus must connect itself to the target cell. Often, this is accomplished by molecules on the virus that mimic cell surface receptors required to interact with other cells. Rhinoviruses, some of which cause the common cold, bind to intercellular adhesion molecules (ICAMs) on the respiratory tract. ICAMs are meant to help white blood cells find their targets, but rhinoviruses have evolved to use them to get into cells. HIV binds to CD4 and CCR5, two surface markers involved in T-cell trafficking. Such binding may restrict the virus to infecting certain species or certain types of cells within a species.

2. Penetration: Once docked, the virus must get its nucleic material inside the cell. Some viruses are taken entirely into the cell. Others

**organelles** membrane bound structures found within a cell

**genes** segments of DNA located on chromosomes that direct protein production

**genomes** the sum of all genes in a set of chromosomes

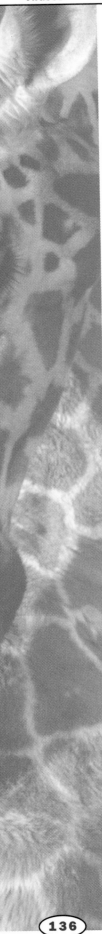

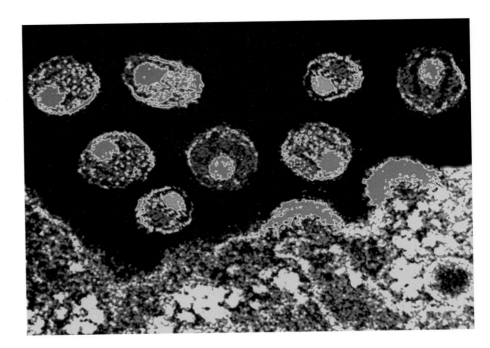

This enhanced photo shows mature HIV viruses.

inject their DNA or RNA through the cell membrane. Still others fuse their membrane with their hosts' and dump the nuclear contents into the cytosol. In order for the virus to use the host's reproduction capabilities, it needs to be inside the cell.

3. Replication: Once the nucleic acids are inside, the virus will use the host's replication machinery to make more copies of itself. Some viruses, such as HIV, are made of RNA instead of DNA. For this reason, these viruses must first transcribe their genomes into DNA in order for cells to copy them. Whether made of DNA or RNA, though, a virus will also force the host cell to make its protein coat and any other proteins necessary to put the virus together. Rather than expending its precious energy on maintenance, the host cell instead is forced to use its energy to make viral parts.

4. Assembly: The various parts of the virus are put together with its freshly copied DNA. Soon the host cell is full of these virions, or individual viral particles, and the virions are ready to infect another cell.

5. Lysis: The virus ruptures the cell and disperses its viral progeny, all off to infect new host cells and begin the cycle again.

Not all viruses lyse (rupture) their host cells. Some may bud off the host cell. Others may become latent and rest in the host's cytosol or even the host's own **chromosomes** for a long time without causing damage. For example, herpes simplex 1 infects individuals and causes cold sores, but does so only intermittently. Cytomegalovirus (CMV) and Epstein-Barr virus (EBV), also known as mononucleosis, both infect the human body and remain latent for life. These viruses are held in check by the immune system and cause no harm, but they never go away completely. Persons with acquired immune deficiency syndrome (AIDS) lose immune function, and CMV and EBV have been known to resurface in persons with AIDS.

**chromosomes** structures in the cell that carry genetic information

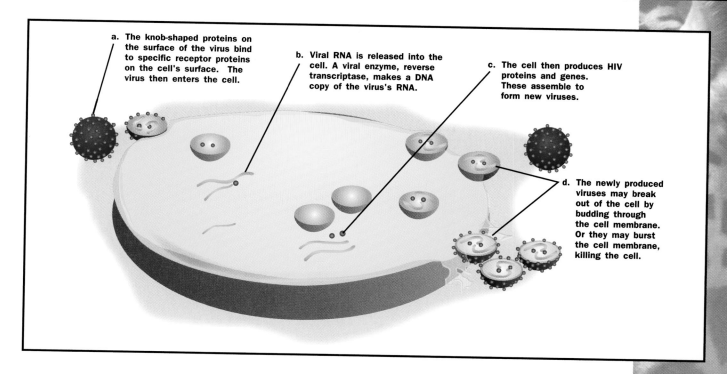

a. The knob-shaped proteins on the surface of the virus bind to specific receptor proteins on the cell's surface. The virus then enters the cell.

b. Viral RNA is released into the cell. A viral enzyme, reverse transcriptase, makes a DNA copy of the virus's RNA.

c. The cell then produces HIV proteins and genes. These assemble to form new viruses.

d. The newly produced viruses may break out of the cell by budding through the cell membrane. Or they may burst the cell membrane, killing the cell.

Virus reproduction. Redrawn from Johnson, 1998.

Because some viruses insert themselves into the host's genome, there is a possibility that they might affect normal gene regulation in the host itself. Viruses can be responsible for certain cellular problems that involve gene regulation. For example, some viruses are thought to be the cause of certain types of cancer. Human papilloma virus, for example, has been associated with cervical cancer. Hepatitis B and hepatitis C cause a majority of the world's cases of liver cancer.

While viruses can be specific for a particular species, cross-species infection happens frequently and sometimes with disastrous results. For example, all fifteen known strains of influenza A virus reside in **aquatic** birds, preferring the intestinal tracts of ducks in particular. As such, fowl fecal matter, as well as seals, whales, pigs, horses, and chickens, have been implicated in a number of human influenza outbreaks.

**aquatic** living in water

As a parasite, the best evolutionary strategy for a virus is for it not to harm the host. It is thought that HIV-1 and HIV-2 were introduced to humanity through the ingestion of uncooked monkeys (a chimpanzee and a sooty mangabey, respectively) sometime in the early twentieth century. These monkeys had been infected with simian immunodeficiency virus (SIV), which is fairly benign to its host. In species such as the mangabeys, African green monkeys, and pigtailed macaques, SIV causes no detectable problems and infection is widespread (it is estimated that some 80 percent of captive sooty mangabeys carry the disease). Infected rhesus monkeys, however, lose their ability to fight disease and waste away. Likewise, when the virus mutated to HIV in humans, it infected human **populations** and continues to cause widespread sickness and death. If HIV is to be with humanity for a long time, then it must become a little less virulent, lest it kill off all of its hosts. SEE ALSO ANIMAL TESTING.

**populations** groups of individuals of one species that live in the same geographic area

*Ian Quigley*

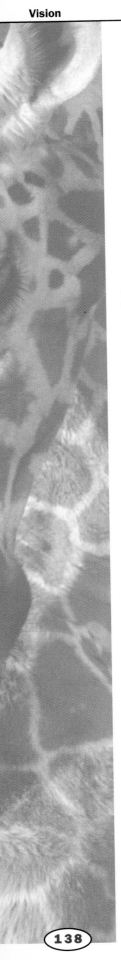

**Bibliography**

Cann, Alan J., ed. *DNA Virus Replication*. New York: Oxford University Press, 2000.

Dalgleish, Angus, and Robin Weiss. *HIV and the New Virus*. San Diego: Academic, 1999.

Johnson, George B. *Biology: Visualizing Life*. New York: Holt, Rinehart and Winston, Inc., 1998.

# Vision

Different levels of vision correlate to the different types of eyes found in various species. The simplest eye receptor is that of planarians, a flatworm that abounds in ponds and streams. Planarians are moderately cephalized and have eye cups (or eyespots) located near the ganglia, a dense cluster of nerve cells with **ventral** nerve cords running along the length of the body.

The receptor cells within the cup are formed by layers of darkly pigmented cells that block light. When light is shined on the cup, stimulation of the photoreceptors occurs only through an opening on one side of the cup where there are no pigmented cells. As the mouth of one eye cup faces left and slightly forward, and the other faces right and forward, the light shining on one side can enter the eye cup only on that side.

This allows the ganglia to compare rates of nerve impulses from the two cups. The planarian will turn until the sensations from the two cups have reached an **equilibrium** and decreased. The observable behavior of the planarian is to turn away from the light source and seek a dark place under an object, an adaptation that protects it from predators.

As evolution progressed and cephalization increased, vison became more complex as well. There are true image-forming eyes of **invertebrates**: compound eyes and **single-lens eyes**. Compound eyes are found in insects and **crustaceans** (phylum Arthropoda) and in some polychaete worms (phylum Annelida). The **compound eye** has up to several thousand light detectors called *ommatidia*, each with its own cornea and lens. This allows each ommatidium to register light from a tiny part of the field of view. Differences in light intensity across the many ommatidia result in a mosaic image.

Although the image is not as sharp as that of a human eye, there is greater acuity at detecting movement—an important adaptation for catching flying insects or to avoid threats of predation. This ability to detect movement is partly due to the rapid recovery of the photoreceptors in the compound eye. Whereas the human eye can distinguish up to 50 flashes per second, the compound eye recovers from excitation rapidly enough to distinguish flashes of at the rate of 330 per second. Compound eyes also allow for excellent color vision, and some insects can even see into the UV range of the spectrum.

The second type of invertebrate eye, the single-lens eye, is found in jellyfish, polychaetes, spiders, and many **mollusks**. Its workings are similar to that of a camera. The single lens focuses light onto the retina, a bilayer of cells that are photosensitive, allowing for an image to be formed.

Vertebrate vision also uses single-lens vision, but it evolved independently and differs from the single-lens eyes of invertebrates. Vertebrate eyes

**ventral** the belly surface of an animal with bilateral symmetry

**equilibrium** a state of balance

**invertebrates** animals without a backbone

**single-lens eyes** an eye that has a single lens for focusing the image

**crustaceans** arthropods with hard shells and jointed bodies and appendages that mainly live in the water

**compound eye** a multi-faceted eye that is made up of thousands of simple eyes

**mollusks** large phylum of invertebrates that have soft, unsegmented bodies and usually have a hard shell and a muscular foot; examples are clams, oysters, mussels, and octopuses

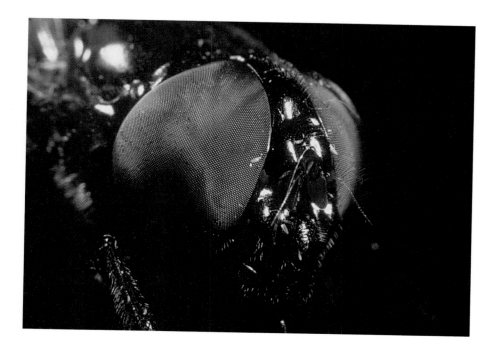

The compound eye of a horsefly.

have the ability to detect an almost countless variety of colors and can form images of objects that are miles away. It can also respond to as little as one photon of light. But as it is actually the brain that "sees," one must also have an understanding of how the eye generates sensations in the form of **action potentials** and how the signals travel to the visual centers of the brain.

## Structure of the Vertebrate Eye

The globe of the eyeball is composed of the *sclera*, a tough, white outer layer of **connective tissue**, and of the *choroid*, a thin, pigmented inner layer. The sclera becomes transparent at the cornea, which is at the front of the eye where light can enter. The **anterior** choroid forms the iris, the colored part of the eye. By changing size, the iris regulates the amount of light entering the pupil, the hole in the center of the iris. Just inside the choroid is the retina, which forms the innermost layer of the eye and contains the photoreceptor cells.

Information from the photoreceptor cells of the retina passes through the optic disc, where the optic nerve attaches to the eye. The optic disc can be thought of as a blind spot in a vertebrate's field of vision because no photoreceptors are present in the disc. Any light that is focused on the lower outside part of the retina, the area of the optic disc, cannot be detected.

The eye is actually composed of two cavities, divided by the lens and the ciliary body. The anterior, smaller cavity is between the lens and the cornea and the **posterior**, larger one is behind the lens, within the eyeball itself.

The ciliary body is involved in constant production of the clear, watery aqueous humor that fills the anterior cavity of the eye. (Blockage of the ducts from which the aqueous humor drains can lead to glaucoma and eventually blindness, as the increased pressure compresses the retina.) The posterior cavity is lubricated by the vitreous humor, a jellylike substance that occupies

**action potential** a rapid change in the electric charge of the cell membrane

**connective tissue** cells that make up bones, blood, ligaments, and tendons

**anterior** referring to the head end of an organism

**posterior** behind or the back

most of the volume of the eye. Both humors function as liquid lenses that help focus light on the retina.

The lens itself is a transparent protein disc that focuses images on the retina. Many fish focus by moving the lens forward and backward, camera-style. In humans and other mammals, focusing is achieved through accommodation, the changing of the shape of the lens. For viewing objects at a distance, the lens is flattened and when viewing objects up close, the lens is almost spherical.

Accommodation is controlled by the ciliary muscle, which contracts to pull the border of the choroid layer of the eye to the lens, causing suspensory ligaments to slacken. There is a decrease in tension and the lens becomes more elastic, allowing the lens to become rounder. For viewing at a distance, the ciliary muscle relaxes, allowing the choroid to expand, thus placing more tension on the suspensory ligament and pulling the lens flatter.

## Signal Transduction

The photoreceptor cells in the retina are of two types: rods and cones. The rods are more sensitive to light but are not involved in distinguishing color. They function in night vision and then only in black and white. Cones require greater amounts of light to be stimulated and are, therefore, not initiated in night vision. However, they are involved in distinguishing colors.

The human retina has approximately 125 million rod cells and approximately 6 million cone cells. The rods and cones account for nearly 70 percent of all receptors in the body, emphasizing the importance of vision in a human's perception of the environment. The numbers of photoreceptors are partly correlated with nocturnal or **diurnal** habits of the species, with nocturnal mammals having a maximum number of rods.

In humans the highest density of rods is at the lateral regions of the retina. Rods are completely absent from the fovea, the center of the visual field. This is why it is harder to see a dim star at night if you look at it directly than if you look at the star at an angle, allowing the starlight to be focused onto rod-populated regions of the retina. However, the sharpest day vision is achieved by looking directly at an object because the cones are most dense in the fovea, approximately 150 thousand cones per square millimeter. Some birds actually have more than one million cones per square millimeter, enabling species such as hawks to spot mice from very high altitudes.

Photoreceptor cells have an outer segment with folded membrane stacks with embedded visual pigments. Retinal is the light-absorbing molecule synthesized from vitamin A and bonded to *opsin*, a membrane protein in the photoreceptor. The opsins vary in structure from one type of photoreceptor to another. The light-absorbing ability of retinal is affected by the specific identity of the opsin partner.

The chemical response of retinal to light triggers a chain of metabolic events, which causes a change in membrane voltage of the photoreceptor cells. The light hyperpolarizes the membrane by decreasing its permeability to sodium ions, so there are fewer **neurotransmitters** being released by the cells in light than in dark. Therefore a decrease in chemical signals to cells with which photoreceptors **synapse** serves as a message that the photoreceptors have been stimulated by light.

**diurnal** active in the daytime

**neurotransmitters** chemical messengers that are released from one nerve cells that cross the synapse and stimulate the next nerve cell

**synapse** the space between nerve cells across which impulses are chemically transmitted

The **axons** of rods and cones synapse with **neurons**, bipolar cells, which synapse with ganglion cells. The horizontal cells and amacrine cells help integrate information before it is transmitted to the brain.

The axons of the ganglion cells form optic nerves that meet at the optic chiasma near the center of the base of the cerebral cortex. The nerve tracts are arranged so that what is in the left field of view of both eyes is transmitted to the right side of the brain (and vice versa).

The signals from the rods and cones follow two pathways: the vertical pathway and the lateral pathway. In the vertical pathway, the information goes directly from receptor cells to the bipolar cells and then to the ganglion cells. In the lateral pathway, the horizontal cells carry signals from one photoreceptor to other receptor cells and several bipolar cells. When the rods or cones stimulate horizontal cells, these in turn stimulate nearby receptors but inhibit more distant receptor and bipolar cells that are not illuminated. This process, termed **lateral inhibition**, sharpens the edges of our field of vision and enhances contrasts in images.

The information received by the brain is highly distorted. Although the anatomy and **physiology** of vision has been extensively studied, there is still much to learn about how the brain can convert a coded set of spots, lines, and movements to perceptions and recognition of objects. SEE ALSO NERVOUS SYSTEM; GROWTH AND DIFFERENTION OF THE NERVOUS SYSTEM.

*Danielle Schnur*

**axons** cytoplasmic extension of a neuron that transmits impulses away from the cell body

**neurons** nerve cells

**lateral inhibition** a phenomenon that amplifies the differences between light and dark

**physiology** the study of the normal function of living things or their parts

### Bibliography

Bradbury, Jack W., and Sandra L. Vehrencamp. *Principles of Animal Communication.* Sunderland, MA: Sinauer, 1998.

Campbell, Neil A., Lawrence G. Mitchell, and Jane B. Reece. *Biology: Concepts and Connections,* 3rd ed. Menlo Park, California: Benjamin/Cummings Publishing Company, 1993.

# Vocalization

Many animals communicate with acoustic signals. Crickets rub a hind leg along a row of protruding spikes; rattlesnakes shake a set of beads in their tail. But vocalization is by far the most common mechanism of acoustic communication. Found only among **vertebrates**, vocalization involves forcing air through a tube across one or more thin membranes located in a tube between the **lungs** and the mouth. These membranes vibrate, producing rapid fluctuations in air pressure in the outgoing air stream that can be detected as sound by a receiver that is tuned to the appropriate frequencies.

Vocalization results from the interaction of four forces acting on each membrane. The first force is air pressure, which is generated by expelling air from a flexible sac, such as a lung. The second force is the elasticity of the membrane, which returns the membrane to its original position after it is disturbed. The third force, which is muscular, forces the membrane into the airflow. The fourth force is called Bernoulli's force, which sucks the membrane into the airflow.

To produce a vocal sound, air is forced through the tube by the contraction of the air sac. Simultaneously, a muscle thrusts the membrane into

**vertebrates** animals with a backbone

**lungs** sac-like, spongy organs where gas exchange takes place

A coyote howls during a snowstorm in Yellowstone National Park.

**anurans** the order of amphibians that contains frogs and toads

Bernoulli's force, also known as Bernoulli's principle, draws its name from its discoverer, Daniel Bernoulli (1700–1782). The principle relates fluid velocity and pressure. Among many other applications, the principle provides an explanation for aircraft lift. The wing shape and the angle of its tip relative to the flow of air are configured so that the air flowing over the wing moves faster than the air under the wing. The pressure difference between the two makes the lift possible.

**habitats** physical locations where an organism lives in an ecosystem

the airflow, causing air pressure to build up. Eventually, this air pressure forces the membrane out of the airflow, rapidly releasing air from the tube. The rapid flow of air past the membrane generates Bernoulli's force, which pulls the membrane back into the airflow.

This cycle repeats until the muscles forcing the membrane into the airflow relax and the elasticity of the membrane pulls it out of the airflow. The rate at which the cycle repeats is equivalent to the frequency of the sound produced and is dependent on the degree of muscular contraction blocking the airflow.

Mammals, **anurans**, and birds each have a unique mechanism of vocalization. All possess a tube called a trachea, which connects the bronchi (tubes leading to the lungs) to the mouth and nose. In mammals, a set of two membranes called the glottis, or vocal cords, partially blocks the airflow through the trachea when a pair of muscles contracts. The diaphragm, an abdominal muscle underneath the lungs and above the intestines, contracts to expand the lungs and relaxes to expel air. Thus, the rate of airflow during vocalization, which can occur only during exhalation, is not under direct muscular control.

The signal produced is periodic and nonsinusoidal, meaning that the rate of airflow varies regularly, but not continuously, around the mean. This produces harmonics, which are sounds at higher multiples of the fundamental frequency. These harmonics can be altered by downstream cavities, such as the mouth, to produce different tones, such as the vowels.

Frogs and toads have two consecutive sets of membranes: the one closer to the lungs acting as the vocal cords, and the farther one referred to as the glottis. The vocal cords control the frequency of the signal, whereas the glottis provides periodic amplitude modulation (AM, as in a radio signal). Expelled air is captured in a throat sac and returned to the lungs to be reused, thereby reducing the amount of work required of the diaphragm. The throat sac also serves as a resonant coupler, its vibrations transferring the signal to the outside air.

Birds have two separate membranes: one between each bronchus and the trachea, together referred to as the syrinx. Each membrane functions independently, allowing birds to produce two different sounds simultaneously. Muscles cause the membranes, which line the side of the syrinx, to buckle, allowing Bernoulli's force to pull them into the airflow.

In addition to lungs, birds possess air sacs throughout the body that are connected to the bronchi and subject to muscular contraction. Thus, exhalation is under direct muscular control, allowing for finely calibrated amplitude modulation. Furthermore, many species employ a labium to control the aperture of the syrinx, and therefore amplitude, onset, and offset of the signal. Finally, the degree of tension in the membranes affects both frequency and amplitude, which are often correlated in bird calls.

Although mammals, anurans, and birds differ in the way they control the frequency and amplitude of their acoustic signals, they share an ability to exploit air as a medium for the rapid transmission of information. This is especially important for animals that live in visually cluttered **habitats** such as trees, or those that communicate over long distances. Sounds do not require localization by the receiver in order to be recognized, whereas visual signals must be noticed first. Therefore, acoustic signals serve well as alarm calls and to attract attention from potential competitors or mates. SEE ALSO ACOUSTIC SIGNALS; COMMUNICATION.

*Brian R. West*

**Bibliography**

Hopp S. L., M. J. Owren, and C. S. Evans. *Animal Acoustic Communication: Sound Analysis and Research Methods.* Berlin: Springer-Verlag, 1997.

# Von Baer's Law

Early developmental stages of a characteristic tend to be more similar among related species than later stages. This means that most characteristics that differentiate **taxa**, and which are commonly used to distinguish among species, represent later modifications to a fundamentally similar developmental plan. Von Baer's Law states that structures that form early in development are more widely distributed among groups of organisms than structures that arise later in development.

**taxa** named taxonomic units at any given level

This law of embryology is named after the nineteenth-century German biologist Karl Ernst von Baer, who first articulated it. Although von Baer was not an evolutionist, his concept of increasing **differentiation** between species during **ontogeny**, or embryonic development, fits well with an evolutionary view of embryology. His law comes from the first two of four generalizations he made in 1828:

1. General features common to a large group of animals appear earlier in the embryo than do specialized features.

2. The development of particular embryonic characters progresses from general to specialized during their ontogeny.

3. Each embryo of a given species, instead of passing through the adult stages of other animals, departs more and more from them as ontogeny progresses (in direct contrast to the invalid biogenetic law of Ernst Haeckel, discussed below).

4. Therefore, the early embryo of a higher animal is never like the adult of a lower animal, only similar to its early embryo.

**differentiation** differences in structure and function of cells in multicellular organisms as the cells become specialized

**ontogeny** the embryonic development of an organism

Von Baer's Law dictates that a generalized state is present in the early embryo. This generalized state gives way to more specialized states as the embryo develop, as can be seen by direct observation of embryos. In other words, an early stage of a chick embryo might be recognizable as a member of the phylum Vertebrata, but not as any particular subtaxon. Later, as structural specialization continues, it is recognizable as a member of the class Aves, and finally as a member of a particular species, *Gallus domesticus.*

Karl Ernst von Baer was the first researcher to explain the law of embryology.

Charles Darwin recognized that von Baer's Law provides a connection between embryology and biological evolution, namely that primitive features tend to be generalized and derived features tend to be specialized. Darwin realized that evolutionary change could be inferred from changes in development.

**homologous** similar but not identical

**phylogenetic** relating to the evolutionary history of a species or group of related species

**notochord** a rod of cartilage that runs down the back of chordates

**vertebrates** animals with a backbone

**dorsal** the back surface of an animal with bilateral symmetry

**mutations** abrupt changes in the genes of an organism

**ontogeny** the embryonic development of an organism

**genetics** the branch of biology that studies heredity

**genes** segments of DNA located on chromosomes that direct protein production

An example of the connection between embryonic development and evolutionary history is the development of the vertebrate limb. Structures of the limb develop in a proximodistal sequence, or from nearest to most distant from the body: shoulder bones develop first, and bones of the digits develop last. Lungfish and coelacanths, distant living relatives of the ancestral tetrapod, have fins with well-developed "shoulders," but their appendages do not contain bones **homologous** to the digits of tetrapods (their similarity in various taxa derives from their common origin in a shared ancestor). Digits appear as specialized, derived features in early tetrapods. Limb morphology, or form, proceeds from generalized to specialized as we climb the **phylogenetic**, or evolutionary, tree, and limb development in the embryos of advanced tetrapods parallels the evolutionary history of these tetrapods.

An additional example of von Baer's Law is seen in the development of the **notocord**. All **vertebrates** develop a **dorsal** cartilaginous rod called the notochord and gill pouches in the pharyngeal, or throat, region. Both are early developmental traits and are common to many species. Later in development, however, these traits are lost and replaced by traits that are distinctive to different groups of vertebrates.

Thus, early embryos of different vertebrate species are remarkably similar. For example, the early embryos of fishes, salamanders, tortoises, chickens, pigs, cows, rabbits, and humans appear similar, yet each species follows its own embryonic developmental trajectory (sequence of events during development) to develop distinctive traits.

**Mutations** that alter early development are usually lethal because they can introduce drastic changes to subsequent development. Because of these developmental constraints, early developmental stages are less likely to change through evolutionary time. These early stages are likely to be similar among taxa and differences among them are more likely to arise later in ontogeny, or embryonic development.

Another nineteenth-century German biologist, Ernst Haeckel, proposed what he thought was a law of evolution. Although Haeckel's law is invalid, it is widely known and resulted in the familiar phrase "ontogeny recapitulates phylogeny." Haeckel's biogenic law thus claims that embryonic development repeats evolutionary history. Embryos, he was saying, repeat the adult stages of ancestral species in chronological order from primitive to most recent and changes are added only during final stages, called terminal addition.

In some cases, the **ontogeny** of a trait does go through stages much like the adult forms of ancestral species. However, although Haeckel's biogenic law describes a possible evolutionary pattern, it provides no causal explanation for the pattern. The biogenetic law eventually lost credibility with the rise of experimental embryology and Mendelian **genetics**. Embryological studies showed that many types of change in developmental timing (heterochrony) are possible and that different parts of the organism may differ in rates of development. Genetic studies have demonstrated that **genes** could effect changes at any stage of development so that terminal addition was not the only possibility. SEE ALSO EMBRYONIC DEVELOPMENT.

*Andrew G. Gluesenkamp*

## Bibliography

Futuyma, Douglas J. *Evolutionary Biology*. Sunderland, MA: Sinauer Associates, Inc., 1986.

Gould, Stephan J. *Ontogeny and Phylogeny*. Cambridge, MA: Belknap Press, 1977.

Raff, Rudolf A. *The Shape of Life*. Chicago: University of Chicago Press, 1996.

Raff, Rudolf A., and Thomas C. Kaufman. *Embryos, Genes, and Evolution: The Developmental-Genetic Basis of Evolutionary Change*. New York: Macmillan, 1983.

# Wallace, Alfred Russel

*Naturalist*

*1823–1913*

Alfred Russel Wallace was born on January 8, 1823, in Usk, Great Britain (Wales). He died at the age of 90 on November 7, 1913. Wallace developed a theory of evolution by **natural selection** independently of Charles Darwin but at nearly the same time. He also founded the field of animal geography, the study of where animals occur on Earth.

As a boy Wallace had a great interest in plants and collected them. In 1844 he began teaching at a boy's school, the Collegiate School in Leicester, Leicestershire, England. There he met the British **naturalist** Henry Walter Bates, who got him interested in insects. In 1848 Wallace and Bates began a four-year expedition to the Amazon. Unfortunately, the ship sank on the return voyage, and most of Wallace's collected specimens were lost. In 1853 he published a book about the journey. In 1854 Wallace began an eight-year tour of the Malay Archipelago in the East Indies. He studied the culture of the native people and the geographical distribution of the animals. He collected and described thousands of new tropical species, and was the first European to see an orangutan in the wild. Wallace discovered that the animals of the Malay Archipelago are divided into two groups of species following an imaginary line, now known as "Wallace's Line." Species west of the line are more closely related to mammals of Asia. Those east of the line are more closely related to mammals of Australia.

While in Malaysia, Wallace thought of the concept of "survival of the fittest" as the key to evolution. In 1858 he wrote about his theory in a paper that he sent to Charles Darwin. Darwin realized that they both had the same revolutionary ideas. The men presented their ideas together in a joint paper to the Linnaean Society in 1858. Darwin is given most of the credit for the theory of evolution by natural selection because he developed the idea in much more detail than Wallace did. Also, Darwin was the person most responsible for its acceptance in the scientific community. Both Wallace and Darwin believed that man evolved to his present bodily form by natural selection. However, Wallace insisted that man's complex mental abilities must have been due to a different, nonbiological force. His activities in spiritual and psychic circles caused many of his fellow scientists to avoid him.

Wallace went on to write many influential books about evolution as well as tales of his journeys. He was an outspoken supporter of socialism, women's

**natural selection** a process by which organisms best suited to their environment are most likely to survive and reproduce

**naturalists** scientists who study nature and the relationships among the organisms

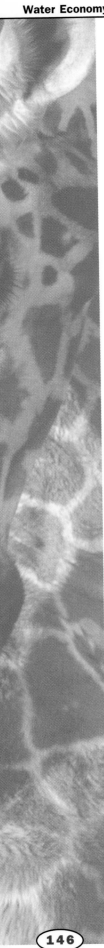

rights, and pacifism. He was awarded the Order of Merit by the British government in 1910.

*Denise Prendergast*

**Bibliography**

Muir, Hazel, ed. *Larousse Dictionary of Scientists*. New York: Larousse Kingfisher Chambers Inc., 1994.

**Internet Resources**

Alfred Russel Wallace. <http://www.wku.edu/%7Esmithch/index1.htm>.

# Water Economy in Desert Organisms

The white-throated wood rat, *Neotoma albigula*, may go its whole life without ever taking a drink of water. It does not need to. The rat obtains all the water it needs from its food. The fact that the rat's diet includes dry seeds and cactus makes this even more surprising. It obtains about half of the water it needs from succulent plants. The white-throated wood rat is also able to chemically synthesize water from the molecules in the food it eats. It shares this adaptation with many other *Neotoma* species, also known as pack rats. The wood rat has many other adaptations that allow it to thrive with little or no water.

The small kit fox, *Vulpes velox*, rarely drinks water. It will if it is thirsty and has the opportunity, but it is generally a long way from the fox's den to any reliable source of water. Therefore, the fox consumes much more prey than it needs for energy requirements. The excess prey is consumed simply for the water it contains. Along with this behavior, the fox has evolved a digestive system that allows it to survive while rarely drinking water.

Both of these are examples of water economy as practiced by desert organisms. Many desert organisms, both plants and animals, have evolved strategies that allow then to survive on little or no water.

## Plant Adaptations

Desert plants have many adaptations that reduce water loss, including extensive root systems to capture as much water as possible and thickened stems to store water. Plants also have waxy or resinous coatings on their leaves and a thick epidermis, and they keep their stomata closed much of the time. However, keeping the stomata closed causes another problem. Since the carbon dioxide molecule is even bigger than the water molecule, anything that reduces water loss also prevents carbon dioxide from entering the leaf. In these plants, the carbon dioxide levels can drop so low that normal **photosynthesis** cannot function.

**photosynthesis** converting sunlight to food

To compensate, many desert plants have evolved a much more efficient form of photosynthesis that uses a molecule with four carbon atoms. Photosynthesis in non-desert plants uses a molecule with three carbon atoms. The two reactions are commonly known as C3 photosynthesis and C4 photosynthesis respectively. The corresponding plants are called C3 plants and C4 plants. Desert plants are frequently C4 plants.

This roadrunner will extract water from its lizard prey. This is one of several adaptations the roadrunner has made in order to exist in an environment with a very limited water supply.

C4 photosynthesis is more efficient at low levels of carbon dioxide. It is more costly in energy terms, but since deserts have plenty of sunlight, low light levels are not a problem.

## Animal Adaptations

Kangaroo rats (*Dipodomys* sp.), small rodents closely related to pocket gophers that are not rats at all, exhibit many of the adaptations common to desert mammals. They have highly efficient kidneys that excrete almost solid urine, thus conserving water. To do this, they have evolved adaptations that allow salt concentration in their urine to be as high as 24 percent, compare to 6 percent in humans.

Also, kangaroo rats have complex passages in their nostrils that condense, collect, and recycle moisture. The temperature at the tips of their noses is 28°C (82°F) while their body temperature is 38°C (100°F). So, as they exhale, the air is cooled and water is condensed and recycled. Humans lose twenty times more water through respiration than do kangaroo rats.

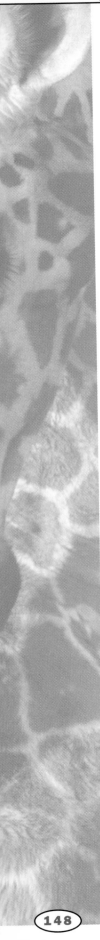

A camel, donkey, and man walk through a desert in Pakistan. Many desert animals have evolved systems for surviving on little or no water.

Kangaroo rats also plug their burrows during the day to maintain a relative humidity in the burrow of 50 percent or higher while the humidity above ground drops to as low as from 10 to 15 percent. Kangaroo rats can also obtain water metabolically from the seeds they eat. Captive kangaroo rats may become stressed enough to drink water, but a wild kangaroo rat may go its entire life without drinking water.

The kangaroo rat is not unique. Many other animals have adaptations that allow them to live in the desert by practicing water economy. Several small rodents share some or all the adaptations of the kangaroo rat. These include the white-throated wood rat (also known as pack rat, *Neotoma* sp.), the canyon mouse (*Peromyscus crinitus*) and many others. Birds also exhibit some adaptations, although most simply escape the dry conditions. The Phainopepla, *Phainopepla nitens*, breeds in the desert in the springtime and escapes to the mountains in summer.

The roadrunner, *Geococcyx californianus*, cannot easily escape the desert, nor does it try. It has several adaptations that allow it to thrive. It reabsorbs water from feces, it excretes excess salt through a nasal gland (other than the roadrunner, an adaptation found only in marine birds), and it is able to utilize effectively the water in its prey. Behavioral modifications include reducing its activity by 50 percent during the heat of midday.

Desert bighorn sheep (*Ovis canadensis nelsoni*) do not drink water in the winter. They obtain all of the water they need from the vegetation they eat. In the summer they must drink about every three days. If water is not available, then the bighorn changes its behavior to rest during daylight hours and feed at night. Since they must rely on plants to supply all their water, they select succulent plants and the tender young pads of prickly pear (after scraping off the thorns with their hooves).

Here are three of the other **physiological** and **behavioral** adaptations for minimizing water loss:

- Many animals are active only at night. Others are crepuscular (active only at dusk and dawn);

- A few desert animals enter a period of dormancy during the heat of summer. These include the round-tailed ground squirrel and several species of desert toads; and

- Birds and reptiles excrete **uric acid** in the form of a semi-solid paste.

Of all the adaptations shared by desert-dwelling animals, the most amazing must be the kangaroo rat's ability to chemically synthesize water. Wild kangaroo rats get by on nothing but dry seeds. They do not drink water even if it is available. So the kangaroo rat is not practicing efficient water economy: it simply does not need to drink water. However, the kangaroo rat must have water in its blood and tissues just like other animals and will die from dehydration if the water in its tissues is deficient. The kangaroo rat gets the water it needs exclusively from digestion of carbohydrates. All animals produce some water, the water of metabolism, this way. But the kangaroo rat metabolizes all the water it ever needs. SEE ALSO DIURNAL; NOCTURNAL.

*Elliot Richmond*

**physiological** the basic activities that occur in the cells and tissues of an animal

**behavioral** relating to actions or a series of actions as a response to stimuli

**uric acid** an insoluble form of nitrogenous waste excreted by many different types of animals

### Bibliography

Burt, William Henry, and Richard Phillip Grossenheider. *The Peterson Field Guide Series: A Field Guide to the Mammals,* edited by R. A. Peterson. Boston: Houghton Mifflin, Company, 1976.

Hoffmeister, Donald Frederick. *Mammals of Arizona.* Tucson: University of Arizona Press, 1986.

Findley, James S., et al. *Mammals of New Mexico.* Albuquerque: University of New Mexico Press, 1975.

Findley, James S. *The Natural History of New Mexican Mammals.* Albuquerque: University of New Mexico Press, 1987.

Krutch, Joseph Wood. *The Voice of the Desert: A Naturalist's Interpretation.* New York: William Sloan Associates, 1962.

MacMahon, James A. *The Audubon Society Nature Guides: Deserts.* New York: Alfred A. Knopf, 1985.

MacMillen, R. E., and E. A. Christopher. "The Water Relations of Two Populations of Non-captive Desert Rodents." In N. F. Hadley ed., *Environmental Physiology of Desert Organisms.* New York: Dowden, Hutchison, and Ross, 1975.

Nowak, Ronald M. *Walker's Mammals of the World,* 5th ed. 2 vols. Baltimore: Johns Hopkins University Press, 1991.

**Whales** *See Morphological Evolution in Whales.*

# Wild Game Manager

A wild game manager oversees wildlife species of game (animals hunted for sport) in their **habitats**, the natural areas where they live. Game includes animals such as deer, elk, wild pigs, duck, geese, and fish. A wild game manager may work for the private sector at sporting clubs or large ranches. Wild game managers may also work for the government (state or federal). They

**habitats** physical locations where an organism lives in an ecosystem

These game wardens tag a white-tailed fawn for population research studies.

**populations** groups of individuals of one species that live in the same geographic area

may be involved in establishing hunting and fishing seasons and regulations. As hunting and fishing on private land become more and more profitable for landowners, there will be more job opportunities in this sector.

Wild game managers make certain that the animals are in good health. They monitor **populations** to see if there are threatening diseases. They may collect population data to ensure that the game exist in sufficient numbers to be safely hunted without being wiped out. Depending on the size of a population, the manager may recommend more or less hunting in a particular season.

Managers study the animals' habitats to see if they are being negatively affected by human activities. This career demands excellent physical condition since the work is outdoors in all kinds of weather. Duties may be difficult, and work hours may be long. A wild game manager may work in remote areas where survival and practical skills are important. The job demands working with people as well as animals. A wild game manager should therefore have good communication skills. Wild game managers should be good at completing tasks on their own and making decisions.

Those who want to work in the field of wild game management can start preparing for this career in high school by taking courses in science, chemistry, math, computers, and English. Outdoor activities such as camping, hunting, fishing, and wildlife photography provide valuable knowledge and experience regarding wildlife and the environment. A college education (bachelor's degree at the minimum) is normally required. In college, potential wild game managers should take courses in the physical and biological sciences, as well as English, statistics, geography, economics, and computers. College training should be supplemented with practical experience. Many students spend their summer months working on a farm or ranch, or working with wildlife, parks, or outdoor recreational agencies.

*Denise Prendergast*

**Bibliography**

Cook, John R., Jr., president, and Kevin Doyle, ed. *The Complete Guide to Environmental Careers in the 21st Century.* Covelo, CA: Island Press, 1999.

**Internet Resources**

*Colorado Division of Wildlife Department of Natural Resources.* <http://www.dnr.state.co.us/wildlife/about/careers.htm>.

# Wildlife Biologist

A career in wildlife biology may consist of conducting research through a university, zoo, or science museum or of working for the government or private corporations. When based out of an academic organization, wildlife biologists dedicate their time to modeling the lives and interactions of particular species or particular environments. Governmental wildlife biologists make recommendations on new legislation that may affect the survival of endangered animals or environments. Wildlife management administrators oversee the running of state and national parks. Governments of countries that strongly rely on **ecotourism** employ wildlife biologists and environmental technicians to create tourist-friendly parks or to reconstruct environments that were damaged in the past. In work for business and corporations, wildlife biologists test the effects of factory pollutants on the environment, advise businesses on where to build new structures, and sometimes actively monitor wildlife preserves located on corporation property.

**ecotourism** tourism that involves travel to areas of ecological or natural interest, usually with a naturalist guide

Wildlife biologists must be able to calculate and predict population sizes, migration patterns, birth and death rates, and environmental interactions, all of which necessitate a strong background in mathematics, physics, and computer science. Most positions offered in this field require a strong background in natural sciences, with a graduate degree in wildlife biology or fishery. Recommended course work includes habitat design, wild bird management, large mammal conservation, wildlife **ecology**, and fisheries ecology. Courses in wild animal veterinary science may also be valuable for some positions. To pursue a career in wildlife biology, it is best to begin early: volunteer at a local zoo, veterinary clinic, or landscaping firm. Seek out wildlife researchers from among college professors and offer to help them with their research. In addition to mathematics and science classes, take

**ecology** the study of how organisms interact with their environment

A group of wildlife biologists examine a tranquilized giant panda.

classes in public speaking, economics, or political science. A bachelor's degree can be followed by graduate studies, and by volunteer or salaried work in a national park.

*Rebecca M. Steinberg*

### Bibliography

Leopold, Aldo. *Game Management.* Boulder, CO: Johnson Books, 2000.

Maynard, Thane, and Jane Goodall. *Working with Wildlife: A Guide to Careers in the Animal World.* New York: Franklin Watts, 1999.

## Wildlife Photographer

Wildlife photography is a loosely-defined profession which demands a passion for nature and art. Wildlife photographers make a career of traveling to remote areas and taking pictures of wild animals and natural scenery.

Wildlife photography began as a hobby of safari hunters, in the early 1900s, who found photographs to be less violent and more permanent reminders of their adventures than the slaying and stuffing of a wild animal. Modern photographers often began as artists, biologists, or park rangers, and taught themselves the technical aspects of the field.

Classes in wildlife photography are offered at colleges and art schools around the world, and many organized safaris and expeditions also provide training and photographic opportunities. Wildlife photographs can be sold to greeting card companies, printed as posters, sold as stock photography, collected into a book, displayed in an art gallery, or sold to a magazine. Ad-

ditionally, some wildlife photographers are hired full-time by nature magazines.

Wildlife photography is a demanding and competitive vocation. Traveling to isolated areas is expensive, as is the photographic equipment necessary for professional quality photos. The work itself is physically and mentally taxing, requiring patience and perfectionism. Because wild animals are wary of humans, wildlife photographers must use a *hide* to get close enough for a clear photograph. Hides are made of anything that conceals form and movement, such as a tiny tent with an opening for the camera, or a car with tinted windows.

Even with a hide, the photographer may have only several seconds in which to take the photo, due to the unpredictability of wild animals, making agility and patience necessary traits. Persistence is also useful. Obtaining special permission to visit national parks in restricted regions or after hours is often essential, although it can be very difficult to accomplish. In addition, the photographer must research the photographic site to understand local lighting conditions, weather, the habits of local

These king penguins pose for a wildlife photographer.

wildlife, and the customs of local peoples. Only the best wildlife photographers earn enough money to support themselves entirely with their photographs.

*Rebecca M. Steinberg*

### Bibliography

Shaw, John. *The Nature Photographer's Complete Guide to Nature Photography.* New York: Garsington, 2000.

Angel, Heather. *Natural Visions: Creative Tips for Wildlife Photography.* New York: Amphoto Books, 2000.

# Wilson, Edward Osborne

## American Biologist
## 1929–

Edward Osborne Wilson was born in Birmingham, Alabama. Wilson originated the field of science called sociobiology, which argues that social animals, including humans, behave mainly according to rules written in their **genes**. Wilson is also considered the world's leading expert on ants.

Wilson's interest in biology began in childhood. He attended the University of Alabama, obtaining a bachelor of science degree in 1949 and a master of science degree in 1950. After obtaining his Ph.D. at Harvard University in 1955, he joined the Harvard faculty. He became a professor in 1964 and curator of entomology (the study of ants) at the university's Museum of Comparative Zoology in 1971.

Wilson has written many important books. In 1971 he published *The Insect Societies*, his definitive work on ants and other social insects. His second major work, *Sociobiology: The New Synthesis* (1975), presented his theories about the biological basis of social behavior. These ideas proved controversial among some scientists and made Wilson famous. His theories caused scientists to discuss and further research the long-standing argument about "nature versus nurture." This is the debate over how much of human behavior is determined by **genetics** and how much by the environment in which a person is raised. Two of Wilson's books won a Pulitzer Prize for general nonfiction: *On Human Nature* in 1979, and *The Ants* (cowritten with Bert Holldobler) in 1991. Wilson's other books include *The Diversity of Life* (1992) and his autobiography, *Naturalist* (1994).

Wilson has made many important scientific discoveries and contributions to biology. He was the first to determine that ants communicate mainly through the exchange of chemical substances called **pheromones**. Wilson worked with the American scientist Robert MacArthur to develop a theory on **populations** of species living on islands. Working with another scientist, W. L. Brown, Wilson developed the concept of **character displacement**. This is the theory that when two closely related species first come into contact, they undergo relatively rapid evolutionary changes. This ensures that they will not have to compete fiercely with one another and that they will not interbreed.

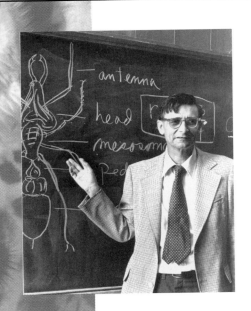

Edward O. Wilson developed the field of sociobiology.

**genes** segments of DNA located on chromosomes that direct protein production

**genetics** the branch of biology that studies heredity

**pheromones** small, volatile chemicals that act as signals between animals that influence physiology or behavior

**populations** groups of individuals of one species that live in the same geographic area

**character displacement** a divergence of overlapping characteristics in two species living in the same environment as a result of resource partitioning

Alarmed by the loss of species throughout the world, Wilson has taken an active role in alerting policymakers and the public about this crisis. Wilson argues that humans are causing the greatest mass extinction of plant and animal species since the extinction of the dinosaurs 65 million years ago. He is an outspoken and active advocate of conserving Earth's resources.

Wilson has received many scientific awards throughout his distinguished career. He was named by Time magazine as one of America's twenty-five most influential people of the twentieth century. He was awarded the National Medal of Science by President Jimmy Carter in 1977. In 1990 he shared Sweden's Crafoord Prize with the American biologist Paul Ehrlich. In 1996, Wilson was named by Time magazine as one of America's twenty-five most influential people of the twentieth century.

*Denise Prendergast*

### Bibliography

Daintith, John, Sarah Mitchell, Elizabeth Tootil, and Derek Jertson. *Biographical Encyclopedia of Scientists*, vol. 2, 2nd ed. Bristol, U.K.: Institute of Physics Publishing, 1994.

Wilson, Edward O. *The Diversity of Life*. Cambridge, MA: Belknap Press, 1992.

### Internet Resources

"Edward Osborne Wilson." *Sociobiology: The New Synthesis.* <http://www.2think.org/sociobiology.shtml>.

# *Xenopus*

The South African clawed frog *Xenopus laevis* is a flat, smooth frog with lidless eyes and webbed feet (in Latin, *xenopus* means "peculiar foot," and *laevis* means "smooth"). The lateral-line system, which consists of sensory hair cells covering the body that are used to detect movements in the water column, persists in adults. The lidless eyes, webbed feet, and maintenance of the lateral-line system in the adult stage are all adaptations to these frogs' lifelong environment. *Xenopus* can live in virtually any amount or quality of water, a necessary adaptation when ponds begin to dry up and become stagnant in the summer. They may also **aestivate** in the mud of dried up ponds, shutting down most of their life processes until more favorable (i.e., wet) conditions arise. These frogs can survive for several months without food or water in aestivation. *Xenopus* are not entirely out of their element on land, either. They have been known to migrate overland in times of drought and to move from overcrowded ponds to colonize new areas at the onset of torrential rains.

*Xenopus* has become a model organism for studies of vertebrate development, and can be found in labs around the world. In fact, escapees have established themselves in a number of wild areas in the Americas. Several characteristics make *Xenopus* amenable to laboratory experimentation. First, they are **aquatic** and can complete their entire life cycle in water. Thus, unlike most amphibians, they can be kept in water tanks instead of **terraria**, which are harder to maintain. Second, they are hardy animals. They eat virtually any type of food and are not prone to disease. Third, they are

**aestivate** a state of lowered metabolism and activity that permits survival during hot and dry conditions

**aquatic** living in water

**terraria** a small enclosure or closed container in which selected living plants and sometimes small land animals, such as turtles and lizards, are kept and observed

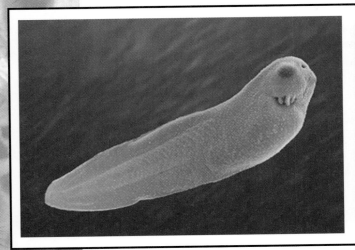 

Comparison of the scanning electron micrograph of the African clawed frog tadpole (left) to the adult frog demonstrates the significant changes undergone during metamorphosis.

**fertilization** the fusion of male and female gametes

**detritus** dead organic matter

**metamorphosis** a drastic change from a larva to an adult

**hormones** chemical signals secreted by glands that travel through the bloodstream to regulate the body's activities

**vertebrates** animals with a backbone

**differentiation** differences in structure and function of cells in multicellular organisms as the cells become specialized

extremely fertile and have a relatively short life cycle. Females can lay up to 1,000 eggs at one spawning, and these eggs can develop into reproductive adults within one year.

The life cycle of *Xenopus*, as in most other frogs, is made up of three main stages: fertilized egg, tadpole, and adult frog. Mating occurs in fresh water with the male clinging to the back of the larger female. **Fertilization** occurs externally—the female lays her eggs in the water while the male releases sperm over them. The tadpoles hatch within three days and start to filter-feed from **detritus** in the water. After about two months, the tadpoles undergo **metamorphosis**, a marked structural transformation in which the legs sprout and the tail is lost. Internally, the skeleton of the head changes shape and teeth appear; and the circulatory, digestive, immune, and other systems are differentiated. Adult *Xenopus* use the claws of their front forelimbs to tear up food which they then manipulate into their tongueless mouths.

*Xenopus* first gained widespread use in labs when researchers discovered that they could be used in human pregnancy tests. The injection of a small amount of urine from a pregnant female under the skin of female *Xenopus* causes them to lay eggs. In fact, an injection of reproductive **hormones** from many vertebrate species causes the same reaction. This proved to be an extremely useful trait, as biologists could now induce egg laying to study the embryology and development of *Xenopus* eggs at any time of year.

Another reason that *Xenopus* are useful in labs is that the large size of their embryos and cells makes visualization of the various developmental stages significantly easier than in most other **vertebrates**. Various staining and labeling techniques are used to follow the cellular and molecular changes occurring in the developing embryo. For example, specific tissues of the embryo may be stained, and then several of these stained embryos may be collected at later life stages. To determine the distribution of this tissue in the embryo, histological sections, or thin slices of embryos from each life stage, are made and observed for staining patterns. In this way, scientists can follow the three-dimensional migration and **differentiation** of a particular tissue type over time.

Scientists use similar techniques to follow spatial and temporal gene **transcription** patterns. For example, to determine when the embryo begins making its own mRNA (and thus its own proteins), rather than relying on maternal mRNA, scientists inject radioactively labeled mRNA **precursors** into developing embryos. As before, sections of the embryos are then made and analyzed for mature radioactive mRNAs. Scientists can also follow the function of a particular gene over time by injecting developing embryos with radioactive probes that bind specifically to mRNA of that gene. This is a very important technique as it lends insight into which **genes** control which developmental processes.

Probably the most striking aspect of *Xenopus* development is metamorphosis. The tadpole and adult stages have fundamentally different **body plans** and life styles. This leads to the question: How are the cells and tissues of the larval stage rearranged and redistributed to form the adult stage, at the same time allowing the developing organism to maintain necessary life functions?

Metamorphosis is initiated by hormones released from the pituitary and thyroid glands. One hormone in particular, thyroxine, affects several tissues and organs of the larvae. In the brain, thyroxine causes most cells to undergo **mitosis**, and the brain becomes larger. However, in the hindbrain, giant Mauthner's cells, which extend down the **spinal chord** and allow the rapid darting movements tadpoles use to avoid predation, degenerate when exposed to thyroxine. Similarly, thyroxine causes the deposition of collagen, a fibrous skin protein, in most areas of the body, but causes the breakdown of collagen in the regressing tail. Thyroxine also causes the initiation of bone development in the limbs.

Thyroxine sets off a cascade of developments in the maturing frog. The **resorption** of the tadpole's tail begins with the resorption of the **notochord** from the tip of the tail to the base. Next the vertebral rudiments and **connective tissue** that surrounded the notochord degenerate. Finally the tail muscles are resorbed and disappear rapidly. This may be because the organism does not want to lose the tadpole form of locomotion (its tail) before the adult form of locomotion (webbed feet) is fully developed. Once the webbed feet are developed, the tail becomes a hindrance and is quickly lost.

The appearance of limbs occurs in the following manner. The hind legs start to bud off just before the front limbs, and shortly thereafter are innervated by new nerve cells. The pelvic girdle (the hip) ossifies first, then the bones of the legs ossify one by one outward from the body. The forelimbs, however, ossify before the shoulder girdle is completed. This is possibly an adaptation which allows filter feeding from the mouth through the gills to continue during development of the forelimbs.

Other than metamorphosis, another unique feature of amphibians such as *Xenopus* is their ability to regenerate their limbs, tail, and many internal organs. If a limb is severed, dead cells at the site are quickly removed through the blood stream. Then nearby **mesodermal** cells, cells with specialized functions in connective or other tissues, multiply and "dedifferentiate" so that they can act as precursors for the new types of cells necessary to form

**transcription** process where enzymes are used to make an RNA copy of a strand of DNA

**precursors** a substance that give rise to a useful substance

**genes** segments of DNA located on chromosomes that direct protein production

**body plan** the overall organization of an animals body

**mitosis** a type of cell division that results in two identical daughter cells from a single parent cell

**spinal cord** a thick, whitish bundle of nerve tissue that extends from the base of the brain to the body

**resorption** absorbing materials that are already in the body

**notochord** a rod of cartilage that runs down the back of Chordates

**connective tissue** cells that make up bones, blood, ligaments, and tendons

**mesoderm** the middle layer of cells in embryonic cells

a complete limb. In general, the ability to regenerate body parts declines over time, especially after metamorphosis is completed.

*Todd A. Schlenke*

**Bibliography**

Bernardini, Giovanni, M. Prati, E. Bonetti, and G. Scari. *Atlas of* Xenopus *Development*. New York: Springer-Verlag, 1999.

Deuchar, Elizabeth M. Xenopus: *The South African Clawed Frog*. New York: John Wiley and Sons, 1975.

Duellman, William E., and Linda Trueb. *Biology of Amphibians*. Baltimore: Johns Hopkins University Press, 1986.

Nieuwkoop, Pieter D., and Jacob Faber, eds. *Normal Table of* Xenopus *Development (Daudin): A Systematical and Chronological Survey of the Development from the Fertilized Egg until the End of Metamorphosis*, 2nd ed. Amsterdam: North-Holland Publishing Company, 1975.

Tinsley, R. C., and H. R. Kobel. *The Biology of* Xenopus. Oxford, U.K.: Oxford University Press, 1996.

# Zoological Parks

As far back as the historical record goes, there is evidence of people keeping wild animals in cages. During the Middle Ages in Europe, rare and exotic animals, and occasionally even foreign natives, were displayed in traveling caravans called menageries. Stationary collections of animals then developed, where the captives were kept in small dark cells, usually alone, with no privacy and nothing to do. They frequently died in a short time and were replaced with new animals captured in the wild. Private zoos on the estates of the wealthy were also popular, with animal dealers supplying birds, reptiles, and mammals from around the world. Often as many as a dozen animals would be killed in captivity or in transport for every one that survived to be sold.

## Protected Environments

**habitat** physical location where an organism lives in an ecosystem

In the late 1800s, a German animal dealer, Carl Hagenbeck, first envisioned the modern zoo. His dream was to create a spacious zoological park where animals could be seen in something resembling their native **habitat**. His park was built in 1907, near Stellingen, Germany, with no fences. Different species such as lions and zebras were kept in the same enclosure, separated by deep moats. Despite commercial popularity, Hagenbeck's ideas did not catch on until the mid-1900s. Animal behaviorists began to emphasize the importance of giving captive animals enough room for some activity and allowing social animals such as monkeys to be in the same cage. They concluded that animals kept in such environments would be healthier, more active, and more interesting to the paying public.

Worldwide, unspoiled habitats began disappearing at an alarming rate by the late twentieth century, as did their inhabitants; more and more wild animals became increasingly rare. Many larger zoos transformed from competing fiefdoms into cooperating members of zoological organizations whose mission became wildlife conservation, research, education of the public, and captive breeding of endangered species.

## Disagreements Over the Best Strategies

As **populations** of some species plummet because of habitat destruction, illegal **poaching**, **pesticides**, and pollution of air and water, conservationists in the United States divide along two lines of thought. The first, which includes the Sierra Club and Friends of the Earth, believes that animals must be protected and saved in their own habitat, that a species is not saved unless it continues to exist in the wild. The second group is led by the U.S. Fish and Wildlife Service, the National Audubon Society, and the American Association of Zoological Parks and Aquariums. They believe that endangered species must be brought into zoos and wildlife parks where new generations can be raised, frequently with much human intervention, and then released back into the wild. Many zoos participate in species survival plans (SSPs), which coordinate breeding efforts for more than fifty species, including cheetahs, rhinoceros, Asian elephants, orangutans, pandas, Puerto Rican crested toads, tamarins, zebras, tapirs, lemurs and Komodo dragons. The long-range goal of the SSP program is to build up large enough populations so that these animals can be returned to their natural habitat. A

Proboscis monkeys at the Bronx Zoo in New York. Many zoos try to incorporate native flora into habitat design.

**populations** groups of individuals of one species that live in the same geographic area

**poaching** hunting game outside of hunting season or by using illegal means

**pesticides** substances that control the spread of harmful or destructive organisms

central computer system, the International Species Inventory System, keeps detailed records of the genetic background of the animals in the program. In order to avoid **inbreeding depression**, which can cause weakened resistance to disease and infertility, animals may be shipped from one part of the world to another to arrange genetically diverse matches. Ultimately, even this tiny gene pool will need to be replenished from the wild to stay vital.

Captive breeding programs have had mixed results. Some species respond well and others, for reasons not yet understood, but probably related to psychological stress, refuse to breed in captivity or, after birth, abandon their offspring.

## New Ideas for Zoological Parks

At the beginning of the twenty-first century, zoos are moving in several directions. As local wild habitats vanish, some small zoos are specializing in local wildlife, providing a glimpse of how indigenous plants and animals interact. Large wildlife parks incorporate hundreds of acres of bogs, woods, meadows, and grasslands and feature native and exotic species that thrive in local conditions. These parks are frequently driven through or have walking trails where people are inobtrusively separated from the animals.

Some zoos have become "bioparks," using aerial walkways, watery moats, and one-way security glass to offer visitors the opportunity to see animals in something resembling natural conditions. These immersion exhibits are sometimes huge re-creations of specific **ecosystems**. Amazonia, at the National Zoo in Washington, D.C., is a 1,400-square-meter (15,000-square-foot) dome filled with thousands of tropical plants, fish, birds, insects, reptiles, amphibians, and mammals. With careful attention to temperature and humidity, it offers a taste of a real tropical jungle. Other immersion exhibits include replicas of Antarctic penguin islands and sandy deserts. Moonlight houses have also become common in many zoos. In these buildings day and night are reversed. **Infrared** lighting allows people to view nocturnal species who rest during the day and are active at night. In large enclosures that have artificial caves and tunnels, bats fly about in search of fruit, nectar, and insects. Desert animals emerge from burrows, coyotes stalk and howl, raccoons hunt for food to wash beside streams, and crocodiles and alligators wallow in marshes before setting off in search of dinner. As people live increasingly urbanized lives, they are looking for zoos to provide a way to get closer to nature and an experience of the wild.

Modern zoos are extremely expensive operations. It is a massive undertaking to meet the daily nutritional needs of hundreds of animals of different species. The food must be as appropriate as possible to each animal and delivered in a way that mimics its natural eating habits. Carnivores need whole animals or large chunks of meat and bone. Primates need to search out fruits, nuts, and seeds that have been hidden in their enclosures. Each elephant requires 45 kilograms (100 pounds) of grass and 14 kilograms (30 pounds) of vegetables a day. Live worms, crickets, guppies, lizards, and mice are all kept on hand for specialized appetites. And some **herbivores** need specific leaves from their homes: acacia for giraffes, eucalyptus for koalas, bamboo for pandas, and hibiscus for leaf-eating monkeys. Abandoned or orphaned infants require precise formulas and extensive care. Some baby birds need to be fed every few minutes to survive.

**inbreeding depression** the loss of fitness due to breeding with close relatives

**ecosystems** self-sustaining collections of organisms and their environments

**infrared** an invisible part of the electromagnetic spectrum where the wavelengths are shorter than red, heat is carried on infrared waves

**herbivores** animals who eat plants only

In nature, sick or injured animals die swiftly. In zoos, each animal represents such a costly investment that most major zoos have full-time medical teams to diagnose and treat disease and provide preventive health care. Tigers and gorillas get dental checks. Females are tested for fertility and carefully monitored during pregnancy. New animals and sick ones are quarantined, and surgeries are performed on broken limbs or tumors.

Naturalistic zoo exhibits attempt to recreate the environment of a given species. This involves scientists, designers, architects, and curators. The exhibit must take into account every detail of each animal's life: breeding, social interaction, exercise, and the food gathering method for which the animal is genetically designed. The insects, plants, and geologic features need to be recreated. Careful regulation of temperature, humidity, and the length and variation of daylight needs to be maintained. The amount of effort involved in recreating even a small ecosystem is beyond the capability of all but the largest zoos.

Although one of the stated goals of zoos is to replenish wild populations, almost no animals have ever been returned to their native habitat. Restoring wild animals is difficult, expensive, and requires the cooperation of the local people to succeed. Craig Hoover, program manager of the Wildlife Trade Monitoring Arm of the World Wildlife Fund, stated that "Zoos have been very successful breeding grounds for many species. But what do you do with those animals when they're not babies anymore? Certainly the open market is the best place to sell them" (Merritt 1991, p. 32). And dozens of major zoos admit to supplying the multibillion dollar a year trade in exotic animals, where rare and endangered species wind up in the private collections of celebrities or as trophy targets on profitable hunting ranches.

Critics of zoos contend that while zoos justify their existence by claiming to educate children, they are teaching the wrong lesson: that it is acceptable to keep wild animals in captivity. Nor is warehousing animals the answer to saving them from extinction. The ultimate salvation of endangered species is in protecting their natural habitats. Perhaps the future of zoos lies in the vision of the Worldlife Center in London, a high-tech zoo with no animals. Visitors observe animals via live satellite links with the Amazon rain forest, the Great Barrier Reef off Australia, and the African savanna and jungle. Rather than putting money into all the infrastructure of a modern zoo, it goes directly to habitat protection to developing nonprofit sanctuaries where hunters and poachers are kept at bay and through which the local people are given an economic incentive to participate in wildlife preservation.

The greatest contribution that zoos may have made is to highlight how little humans understand the incredibly intricate mechanisms of life on Earth. Human technology, while capable of destroying vast ecosystems, is insufficient to create and maintain them. Zoos began as a symbol of human conquest of wild animals. Perhaps modern zoos can serve as sanctuaries for animals rescued from ill treatment and as reminders to be more respectful of nature and the place of humans in it. SEE ALSO HABITAT LOSS; HABITAT RESTORATION; NATURAL SELECTION; POPULATIONS; SELECTIVE BREEDING.

*Nancy Weaver*

The Zoological Society of San Diego has developed a different and engaging interface for accessing information on its web site. This new site is titled "E-Zoo," and is available at http://www.sandiegozoo.org/virtualzoo/homepage.html. Between the interactive games, the videos, and the e-postcards, the Zoological Society has helped to make learning about animals fun from any location—even your desk chair.

**Bibliography**

Challindor, David, and Michael H. Robinson. *Zoo Animals: A Smithsonian Guide.* New York: Macmillan, 1995.

Merritt, Clifton. "Low Class, No Class Scumballs." *Animals Agenda* 11, no. 10 (1991):32.

# Zoologist

Zoology is a branch of biology that involves the study of animals. Zoology is a very broad field; zoologists study subjects ranging from single-celled organisms to the behavior of groups of animals. Some zoologists study the biology of a particular animal, others study the structure and function of the bodies of animals. Some zoologists are interested in **heredity**, how the characteristics of animals are passed from one generation to the next. Others study the way animals interact with other animals or their surrounding environment. Many zoologists are classified by the animal they study. For example, herpetologists study reptiles and amphibians. Entomologists specialize in insects. Ornithologists study birds. Mammalogists specialize in mammals.

Many zoologists are involved in research and development. Some conduct basic research to expand knowledge about animals. Others may conduct applied research, which is used to directly benefit humans. Applied research may be used to develop new medicines, make livestock more resistant to disease, control pests, or help the environment. Researchers may be employed at universities, government agencies, nonprofit organizations, scientific institutions, or private industries. Many zoologists work at zoos, aquariums, and museums. Research can involve fieldwork, laboratory work, and writing up the results for publication. Most zoologists spend only two to eight weeks in the field each year. Junior scientists spend more time in the field than senior scientists, observing animals and collecting data. Senior scientists spend much of their time coordinating research, overseeing junior staff, and writing grant proposals, or obtaining funds in some other way. Zoologists involved in research need at least a bachelor's degree. Advanced degrees (master's and Ph.D.) can also be very helpful.

The following types of companies/organizations may employ zoologists:

- animal hospitals
- pet stores
- food companies
- biotechnology firms
- humane societies
- pharmaceutical companies
- chemical companies
- medical laboratories
- clinics/hospitals
- research laboratories
- veterinarian schools

**heredity** the passing on of characteristics from parents to offspring

This zoologist weighs a polar bear cub.

- national parks
- dog-training schools
- natural history museums
- environmental companies
- veterinarian-supply houses
- pest-control agencies
- government agencies

Some zoologists are primarily teachers. They can teach at the high-school level or at the university/college level. Teaching at the high-school level requires a bachelor's degree and state certification. Teaching at the university level generally requires a doctoral (Ph.D.) degree.

At the high-school level, persons interested in becoming zoologists should study mathematics, chemistry, physics, biology, English, writing, and computer studies. In college, persons obtaining a bachelor's degree in zoology will study these same types of subjects. In graduate school, a student specializes in a particular area of interest. SEE ALSO ANIMAL.

*Denise Prendergast*

**Bibliography**

Cosgrove, Holli R., ed. *Encyclopedia of Careers and Vocational Guidance*, 11th ed. Chicago: Ferguson Publishing Company, 2000.

# Zooplankton

The word "plankton" refers to the floating marine organisms that live on the surface of oceans. These organisms can be plants or animals. The plant forms are microscopic algae whose photosynthesis reactions provide the Earth's atmosphere with the majority of its oxygen. The other type of plankton, composed of tiny animals, is called zooplankton.

Zooplankton is made up of hundreds of thousands of different species of animals. Some are baby or larval forms of the animals while others spend their whole life as free-floating organisms. The entire scope of species of zooplankton is enough for scientists to have identified whole communities of these organisms. These communities are very dynamic in that they change their structure and **populations** on a seasonal basis.

Many members of a zooplankton community begin their lives in **estuaries** where crabs, fishes, and a whole host of various **invertebrates** come to breed. The calm and relatively shallow waters of an estuary provide a safe place for eggs to survive and hatch. Upon hatching, the tiny larvae are too small to succumb to the effects of gravity in the water and so begin their journeys as floating animals.

In many species of zooplankton, the larval forms look nothing like the adults they will become. A remarkable example is the flounder fish. It starts its life as a small larval floating form that looks very much like a common fish. It drifts in the water for about forty or forty-five days until it begins its transformation into a bottom-dwelling flat adult. While drifting in the

**populations** groups of individuals of one species that live in the same geographic area

**estuaries** an area of brackish water where a river meets the ocean

**invertebrates** animals without a backbone

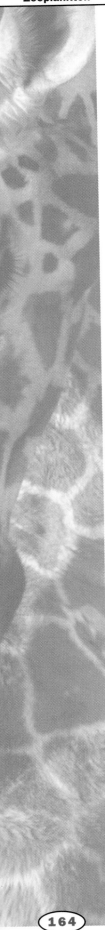

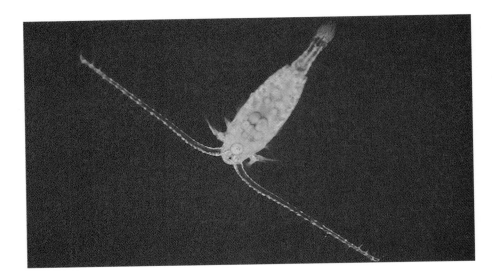

A living crustacea copepoda.

ocean it feeds on other plankton and begins to grow until it begins its juvenile phase. As a juvenile, it drifts to the ocean bottom and flattens out, with one of its eyes rotating around its head to sit next to the other eye. As a bottom-dwelling adult it becomes a flat fish with eyes on the upper surface of its body. Then it is ready to produce more planktonic larvae.

Many members of the zooplankton community feed on other members of the population, and in turn become the meals of other larger predators. Eventually, the whole zooplankton community becomes the bottom of a food chain for an entire food web stretching from the smallest fish to the largest whale. Many of the ocean's largest animals feed on zooplankton. Many whales have feeding structures called **baleen** that filter the zooplankton from the water. In the polar regions, a small component of the zooplankton community called **krill** is the basic diet of the many summer-feeding whales.

One of the benefits of becoming a pelagic, or open ocean-dwelling, organism for a specific population is that the drifting currents move the offspring from one place to another. This ensures species distribution, which is critical to the survival of many species. It keeps genetic diversity high and populations healthy.

The waterborne distribution of zooplankton helps its population survive harsh environmental conditions such as freezing, high heat, large storms, and other severe natural phenomena. Because riding the ocean currents distributes many species worldwide, only small portions of a population may be seriously affected by these conditions.

For many planktonic forms, their lifestyleas organisms suspended on the ocean surface, means that they can avoid unfavorable conditions in the deeper regions of the water column. Those species that eventually drift to the bottom and complete their **metamorphosis** avoid some forms of predation until they are of a suitable size or form to avoid attack.

One of the concerns raised by the increasing depletion of the ozone layer is how the increased influx of ultraviolet radiation that it causes will affect zooplankton. Because they are the basis of the food chain for a great many animals in lakes and oceans, many scientists and others are concerned that the tiny covering around these larval animals may not be

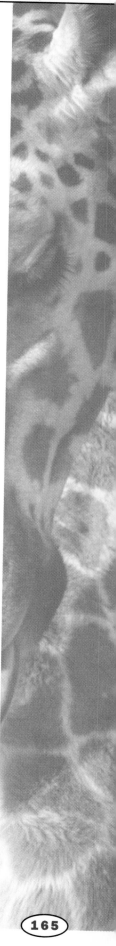

strong enough to withstand the impact of increased radiation. As in all food web chains, the zooplankton provide a foundation for so many other and larger food species that some forms of higher predators may be seriously impacted.

Until the increasing ozone loss is curbed, the protection of estuaries, deltas, and other coastal planktonic breeding grounds is crucial for the continued production of zooplankton. More and more, scientists are discovering how important zooplankton are for the health of the marine ecosystems. SEE ALSO FOOD WEB; PLANKTON.

*Brook Ellen Hall*

### Bibliography

Pechnik, J. *Biology of the Invertebrates*. Boston: Prindle, Weber, & Schmidt, 1985.

Nybakken, J. *Marine Biology: an Ecological Approach*. New York: Harper & Row, 1988.

Moyle, P., and J. Chech Jr. *Fishes: An Introduction to Icthyology*. Upper Saddle River, NJ: Prentice Hall, 1996.

# Photo and Line Art Credits

The illustrations and tables featured in Animal Sciences were created by GGS Information Services. The photographs appearing in the text were reproduced by permission of the following sources:

## Volume One

*Acoustic Signals* (**4**): Zig Leszczynski/Animals Animals; *Adaptation* (**7**): JLM Visuals, Ken Cole/Animals Animals; *African Cichlid Fish* (**9**): Gerard Lacz/Animals Animals; *Aggression* (**11**): Lynn Stone/Animals Animals; *Agnatha* (**14**): Omikron, National Audobon Society Collection/Photo Researchers, Inc.; *Allometry* (**16**): Redrawn from Scott Gilbert; *Allometry* (**15**): Redrawn from Scott Gilbert; *Altruism* (**18**): Robert J. Huffman/Field Mark Publications; *Amphibia* (**20**): Robert J. Huffman/ Field Mark Publications; *Animal Rights* (**25**): AP/Wide World Photos; *Animal Testing* (**29**): Yann Arthus-Bertrand/Corbis; *Annelida* (**31**): E. R. Degginger/Animals Animals; *Antibody* (**34**): Redrawn from Hans and Cassady; *Antlers and Horns* (**35**): Kennan Ward/Corbis; *Apiculture* (**37**): Breck P. Kent/Animals Animals; *Aposematism* (**39**): R. Andrew Odum, Toledo Zoological Society; *Aquaculture* (**42**): David Samuel Robbins/Corbis; *Aristotle* (**45**): UPI/Bettmann Newsphotos/Corbis; *Arthropoda* (**47**): Dr. Dennis Kunkel/ Phototake; *Arthropoda* (**50**): JLM Visuals; *Aves* (**53**): Joe McDonald/Animals Animals; *Bailey, Florence Augusta Merriam* (**57**): Courtesy of the Library of Congress; *Bates, Henry Walter* (**58**): Courtesy of the Library of Congress; *Behavior* (**60**): Ross, Edward S., Dr.; *Behavior* (**61**): Carl Purcell/Corbis; *Binomial (Linnaean System)* (**64**): Archive Photos, Inc.; *Biodiversity* (**68**): Joerg Sarback/AP/Wide World Photo;

*Bioethics* (**71**): Jacques M. Chenet/Corbis; *Bioethics* (**70**): Robyn Beck/Corbis; *Biological Evolution* (**75**): Sally Morgan/Ecoscene/Corbis; *Biological Evolution* (**76**): P. Parks-OSF/ Animals Animals; *Biological Pest Control* (**80**): Bill Beatty/Animals Animals; *Biomechanics* (**84**): Brad Nelson/Phototake; *Biomes* (**88**): W. Wayne Lockwood, M.D./Corbis; *Biomes* (**89**): Robert J. Huffman/Field Mark Publications; *Blood* (**98**): Richard T. Nowitz/Phototake; *Body Cavities* (**100**): Redrawn from Holt, Rinehart and Winston; *Body Cavities* (**101**): Nancy Sefton, National Audobon Society Collection/ Photo Researchers, Inc.; *Body Plan* (**108**): JLM Visuals; *Bone* (**111**): CNRI/Phototake; *Burgess Shale and Ediacaran Faunas* (**113**): Lowell Laudon/JLM Visuals; *Cambrian Explosion* (**117**): Robert J. Huffman/Field Mark Publications; *Cambrian Period* (**119**): Mickey Gibson/Earth Scenes; *Camouflage* (**120**): Art Wolfe/The National Audubon Society Collection/Photo Researchers, Inc.; *Cancer* (**123**): Luis M. De La Maza/Phototake; *Carson, Rachel* (**128**): UPI/Bettmann Newsphotos/Corbis; *Cartilage* (**129**): Dr. Mary Notter/Phototake; *Cell Division* (**134**): Redrawn from Gale Research; *Cells* (**136**): Redrawn from New York Public Library Desk Reference; *Cephalization* (**139**): P. Parks-OSF/ Animals Animals; *Cestoda* (**142**): Custom Medical Stock Photo, Inc.; *Chitin* (**143**): Chris McLaughlin/Animals Animals; *Chondricthyes* (**145**): Joyce and Frank Burek/Animals Animals; *Chordata* (**147**): Ralph White/Corbis; *Circadian Rhythm* (**150**): George McCarthy/ Corbis; *Circulatory system* (**152**): GCA/CNRI/ Phototake; *Cnidaria* (**155**): Nancy Rotenberg/ Animals Animals; *Coevolution* (**161**): Robert J.

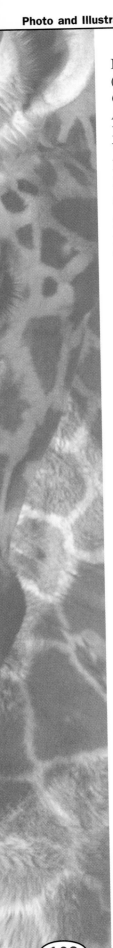

Huffman/Field Mark Publications; *Coevolution* (**159**): Michael and Patricia Fogden/Corbis; *Communication* (**165**): James E. Lloyd/Animals Animals; *Community Ecology* (**168**): Richard Baetson/U.S. Fish and Wildlife Service, Washington D.C.; *Competition* (**174**): Ross, Edward S., Dr.; *Conservation Biology* (**177**): Fred Whitehead/Animals Animals; *Constraints on Animal Development* (**179**): Robert Lubeck/Animals Animals; *Convergence* (**185**): E. R. Degginger/Animals Animals; *Courtship* (**186**): Michael Fogden/Animals Animals.

## Volume Two

*Cretaceous* (**2**): Richard P. Jacobs/JLM Visuals; *Cultures and Animals* (**6**): Duluth Convention and Visitor's Bureau; *Cultures And Animals* (**8**): Craig Lovell/Corbis; *Darwin, Charles* (**10**): Courtesy of the Library of Congress; *DDT* (**12**): AP/Wide World Photos; *Defense* (**14**): Mark N. Boulton/The National Audubon Society Collection/Photo Researchers, Inc.; *Diamond, Jared* (**18**): AP/Wide World Photos; *Digestive System* (**22**): Roger De La Harpe/Animals Animals; *Dinosaurs* (**25**): Graves Museum of Archaeology and Natural History/AP/Wide World Photos; *Domestic Animals* (**29**): Tim Thompson/Corbis; *Dominance Hierarchy* (**32**): Tim Wright/Corbis; *Drosophila* (**34**): UPI/Bettmann Newsphotos/Corbis; *Echinodermata* (**36**): Herb Segars/Animals Animals; *Ecology* (**42**): U.S. National Aeronautics and Space Administration (NASA); *Ecosystem* (**47**): James Watt/Animals Animals; *Elton, Charles Sutherland* (): Animals Animals/©; Scott Johnson; *Embryology* (**52**): Science Pictures Limited/Corbis; *Embryonic Development* (**54**): Redrawn from Scott Gilbert; *Embryonic Development* (**59**): Redrawn from Scott Gilbert; *Endangered Species* (**62**): Vince Streano/The Stock Market; *Entomology* (**72**): Doug Wechsler/Earth Scenes; *Environmental Degradation* (**74**): George H. H. Huey/Earth Scenes; *Environmental Impact* (**78**): David Barron/Earth Scenes; *Excretory and Reproductive Systems* (**88**): Joe McDonald/Animals Animals; *Exotic Species* (**91**): C. C. Lockwood/Animals Animals; *Extinction* (**100**): Glen Smart/U.S. Fish and Wildlife Service/Washington, D.C.; *Extremophile* (**103**): TC

Nature/Animals Animals; *Farming* (**105**): Tim Page/Corbis; *Feeding* (**111**): Malcolm S. Kirk/Peter Arnold, Inc.; *Feeding Strategies* (**114**): Bruce Davidson/Animals Animals; *Fertilization* (**117**): Hank Morgan/National Audobon Society Collection/Photo Researchers, Inc.; *Flight* (**122**): Zefa Germany/The Stock Market; *Foraging Strategies* (**130**): Suzanne Danegger/Photo Researchers, Inc.; *Fossey, Dian* (**131**): AP/Wide World Photos; *Fossil Fuels* (**133**): H. David Seawell/Corbis; *Functional Morphology* (**137**): Tim Davis/Photo Researchers; *Genes* (**140**): Biophoto Associates/National Audobon Society Collection/Photo Researchers, Inc.; *Genetic Engineering* (**149**): Lowell Georgia/Corbis; *Genetic Variation In A Population* (**153**): Eastcott-Momatiuk/Animals Animals; *Genetically Engineered Foods* (**156**): Lowell Georgia/Corbis; *Geneticist* (**157**): Photo Researchers, Inc.; *Genetics* (**160**): AP/Wide World Photos; *Gills* (**164**): Breck P. Kent/Animals Animals; *Global Warming* (**168**): Bernard Edmaier/Science Photo Library/Photo Researchers, Inc.; *Goodall, Jane* (**171**): AP/Wide World Photos; *Gould, Steven Jay* (**172**): Wally McNamee/Corbis; *Habitat* (**178**): Johnny Johnson/Animals Animals; *Habitat Loss* (**180**): Ecoscene/Corbis; *Habitat Loss* (**182**): Francois Gohier/National Audobon Society Collection/Photo Researchers, Inc.

## Volume Three

*Habitat Restoration* (**3**): Eastcott-Momatiuk/Earth Scenes; *Haeckel's Law Of Recapitulation* (**5**): Courtesy of the Library of Congress; *Haldane, J. B. S.* (**6**): Courtesy of the Library of Congress; *Heterochrony* (**9**): Christian Testu/BIOS; *Home Range* (**11**): Darren Bennet/Animals Animals; *Hormones* (**17**): Barbra Leigh/Corbis; *Horse Trainer* (**19**): Jerry Cooke/Animals Animals; *Human Commensal and Mutual Organisms* (**31**): LSF OSF/Animals Animals; *Human Evolution* (**35**): John Reader/Science Photo Library/Photo Researchers, Inc.; *Human Populations* (**41**): AP/Wide World Photos; *Human–Animal Conflicts* (**28**): John Nees/Animals Animals; *HunterGatherers* (**44**): Anthone Bannister/Earth Scenes; *Hunting* (**46**): Raymond Gehman/Corbis; *Imprinting* (**50**): Robert J. Huffman/Field Mark

Publications; *Instinct* (**52**): Anup Shah/Animals Animals; *Interspecies Interactions* (**55**): JLM Visuals; *Iteroparity And Semelparity* (**58**): Kennan Ward/The Stock Market; *Jurassic* (**60**): E. R. Degginger/Earth Scenes; *Keratin* (**63**): L. Lauber-OSF/Animals Animals; *Lamarck* (**69**): Courtesy of the Library of Congress; *Leakey, Louis And Mary* (**70**): UPI/Bettmann Newsphotos/Corbis; *Learning* (**72**): Erwin and Peggy Bauer/Animals Animals; *LeviMontalcini, Rita* (**74**): AP/Wide World Photos; *Linnaeus, Carolus* (**77**): U.S. National Library of Science; *Living Fossils* (**80**): Tom McHugh/Steinhart Aquarium/ Photo Researchers, Inc.; *Locomotion* (**88**): Jeffrey L. Rotman/Corbis, Richard Alan Wood/Animals Animals, Ross, Edward S., Dr., Robert Winslow/Animals Animals; *Lorenz, Konrad* (**91**): UPI/Bettmann Newsphotos/Corbis; *Malaria* (**95**): Redrawn from Hans and Cassady; *Malthus, Thomas Robert* (**98**): Archive Photos, Inc.; *Mammalia* (**103**): Leonard Lee Rue III/Animals Animals; *Mayr, Ernst* (**106**): AP/Wide World Photos; *Mendel, Gregor* (**108**): Archive Photos, Inc.; *Mesenchyme* (**110**): Lester V. Bergman/Corbis; *Migration* (**117**): AP/Wide World Photos; *Mimicry* (**122**): Breck P. Kent/Animals Animals; *Molecular Biologist* (**126**): Philippe Plailly/National Audubon Society Collection/ Photo Researchers, Inc.; *Molting* (**135**): Ross, Edward S., Dr.; *Morphological Evolution in Whales* (**137**): Bomford/T.Borrill-OSF/ Animals Animals; *Mouth, Pharynx, Teeth* (**141**): H. Pooley/OSF/Animals Animals; *Muscular System* (**143**): Custom Medical Stock Photo, Inc.; *Museum Curator* (**145**): Robert J. Huffman/Field Mark Publications; *Natural Resources* (**147**): Richard Hamilton Smith/Corbis; *Nematoda* (**151**): Eric V. Grave/ Photo Researchers, Inc.; *Nervous System* (**157**): Redrawn from Neil Campbell.; *Nervous System* (**155**): Secchi-Lecague/Roussel-UCLAF/ CNRI/Science Photo Library/National Audubon Society Collection/Photo Researchers, Inc.; *Osteichthyes* (**169**): A. Kuiter/OSF/ Animals Animals; *Parasitism* (**179**): Frank Lane Picture Agency/Corbis; *Pasteur, Louis* (**180**): Courtesy of the Library of Congress; *Peppered Moth* (**186**): Breck P. Kent/Animals Animals.

## Volume 4

*Permian* (**2**): Richard P. Jacobs/JLM Visuals; *Pesticide* (**4**): Arthur Gloor/Earth Scenes; *Phylogenetic Relationships of Major Groups* (**6**): Redrawn from Caroline Ladhani.; *Plankton* (**12**): Douglas P. Wilson/Frank Lane Picture Agency/Corbis; *Pleistocene* (**16**): Breck P. Kent/Earth Scenes; *Pollution* (**21**): Wesley Bocxe/National Audubon Society Collection/ Photo Researchers, Inc.; *Population Dynamics* (**26**): Wolfgang Kaehler/Corbis; *Populations* (**32**): JLM Visuals; *Predation* (**37**): Johnny Johnson/Animals Animals; *Primates* (**40**): Barbara Vonn Hoffmann/Animals Animals; *Reproduction, Asexual and Sexual* (): Animals Animals/©; Zig Leszczynski; *Reptilia* (**52**): Animals Animals/©; Zig Leszczynski; *Respiration* (**55**): Ron Boardman, Frank Lane Picture Agency/Corbis; *Respiratory System* (**58**): Photo Researchers, Inc.; *Rotifera* (**60**): P. Parks/OSF/Animals Animals; *Scales, Feathers, and Hair* (**62**): Dr. Dennis Kunkel/Phototake; *Selective Breeding* (**65**): George Bernard/ Animals Animals; *Sense Organs* (**68**): Robert J. Huffman/Field Mark Publications; *Sexual Dimorphism* (**74**): Gregory G. Dimijian/Photo Researchers, Inc.; *Sexual Selection* (**77**): Animals Animals/©; Darren Bennett; *Shells* (**81**): Spreitzer, A.E., Mr.; *Simpson, George Gaylord* (**85**): Courtesy of the Library of Congress; *Skeletons* (**86**): E. R. Degginger/ Animals Animals; *Social Animals* (**91**): G. I. Bernard/OSF/Animals Animals; *Sociality* (**95**): JLM Visuals;*Stevens, Nettie Maria* (**99**): Science Photo Library/Photo Researchers, Inc.; *Sustainable Agriculture* (**101**): Michael Gadomski/Earth Scenes; *Territoriality* (**106**): Jack Wilburn/Animals Animals; *Threatened Species* (**113**): Kenneth W. Fink/Photo Researchers, Inc.; *Tool Use* (**115**): C. Bromhall/ OSF/Animals Animals; *Trematoda* (**123**): Photo Researchers, Inc.; *Triassic* (**125**): Robert J. Huffmann/Field Mark Publications; *Urochordata* (**130**): Joyce and Frank Burek/ Animals Animals; *Vertebrata* (**132**): Breck P. Kent/JLM Visuals; *Veterinarian* (**134**): Peter Weiman/Animals Animals; *Viruses* (**136**): Scott Camazinr/National Audubon Society Collection/Photo Researchers, Inc.; *Vision* (**139**): J. A. L. Cooke/OSF/Animals Animals;

# Glossary

**abiogenic:** pertaining to a nonliving origin

**abiotic:** nonliving parts of the environment

**abiotic factors:** pertaining to nonliving environmental factors such as temperature, water, and nutrients

**absorption:** the movement of water and nutrients

**acid rain:** acidic precipitation in the form of rain

**acidic:** having the properties of an acid

**acoelomate:** an animal without a body cavity

**acoelomates:** animals without a body cavity

**acoustics:** a science that deals with the production, control, transmission, reception, and effects of sound

**actin:** a protein in muscle cells that works with myosin in muscle contractions

**action potential:** a rapid change in the electric charge of the cell membrane

**active transport:** a process requiring energy where materials are moved from an area of lower to an area of higher concentration

**adaptive radiation:** a type of divergent evolution where an ancestral species can evolve into an array of species that are specialized to fit different niches

**adenosine triphoshate:** an energy-storing molecule that releases energy when one of the phosphate bonds is broken; often referred to as ATP

**aestivate:** a state of lowered metabolism and activity that permits survival during hot and dry conditions

**agnostic behavior:** a type of behavior involving a contest of some kind that determines which competitor gains access to some resource such as food or mates

**alkaline:** having the properties of a base

**allele:** one of two or more alternate forms of a gene

**alleles:** two or more alternate forms of a gene

**allometry:** relative growth of one part of an organism with reference to another part

**allopatry:** populations separated by a barrier

**alluvial:** sediments from flowing water such as silt, sand, mud, and gravel

**alpha:** the dominant member of a group

**altruistic behavior:** the aiding of another individual at one's own risk or expense

**alveoli:** thin-walled sacs in the lungs where blood in capillaries and air in the lungs exchange gases

**ameloblasts:** cells that form dental enamel

**amiote:** embryo of a vertebrate that is surrounded by a fluid-filled sac

**ammonites:** an extinct group of cephalopods with a curled shell

**amnion:** the membrane that forms a sac around an embryo

**amniote:** a vertebrate which has a fluid-filled sac that surrounds the embryo

**amniotes:** vertebrates which have a fluid-filled sac that surrounds the embryo

**anadromous:** moving from the ocean up a river to spawn

**analogous:** a similarity in structures between two species that are not closely related

**anemia:** a condition that results from a decreased number of red blood cells

**angiosperms:** a flowering plant that produces seeds within an ovary

**annelids:** segmented worms

**anoxic:** an environment that lacks oxygen

**anterior:** referring to the head end of an organism

**anterior pituitary:** the front part of the pituitary gland that produces hormones that stimulate the production of sperm and testosterone in the testes

**antibodies:** proteins in the plasma produced by B cells and plasma cells in reaction to foreign substances or antigens

**antigen:** foreign substances that stimulate the production of antibodies in the blood

**anurans:** the order of amphibians that contains frogs and toads

**aphrodisiac:** a substance or object that is thought to arouse sexual desire

**aphrodisiacs:** substances or objects that are thought to arouse sexual desire

**aposematic:** a feature or signal that serves to warn

**aposematic coloration:** a bright coloration in animals with physical or chemical defenses that act as a warning to predators

**appendicular:** having to do with arms and legs

**appendicular skeleton:** part of the skeleton with the arms and legs

**aquatic:** living in water

**aragonite:** a mineral form of calcium carbonate

**arboreal:** living in trees

**Archae:** an ancient lineage of prokaryotes that live in extreme environments

**arthropod:** a phylum of invertebrates characterized by segmented bodies and jointed appendages such as antennae and legs

**arthropods:** members of the phylum of invertebrates characterized by segmented bodies and jointed appendages such as antennae and legs

**artificial pollination:** manual pollination methods

**asexual reproduction:** a reproduction method with only one parent, resulting in offspring that are genetically identical to the parent

**asymmetrical:** lacking symmetry, having an irregular shape

**aural:** related to hearing

**autonomic nervous system:** division of the nervous system that carries nerve impulses to muscles and glands

**autotroph:** an organism that makes its own food

**autotrophs:** organisms that make their own food

**axial skeleton:** the skeleton that makes up the head and trunk

**axon:** cytoplasmic extension of a neuron that transmits impulses away from the cell body

**axons:** cytoplasmic extensions of a neuron that transmit impulses away from the cell body

**B-lymphocytes:** specialized cells produced from stem cells in the bone marrow that secrete antibodies that bind with antigens to form a pathogen fighting complex

**bacterium:** a member of a large group of single-celled prokaryotes

**baleen:** fringed filter plates that hang from the roof of a whale's mouth

**Batesian mimicry:** a type of mimicry in which a harmless species looks like a different species that is poisonous or otherwise harmful to predators

**behavioral:** relating to actions or a series of actions as a response to stimuli

**benthic:** living at the bottom of a water environment

**bilateral symmetry:** characteristic of an animal that can be separated into two identical mirror image halves

**bilaterally symmetrical:** describes an animal that can be separated into two identical mirror image halves

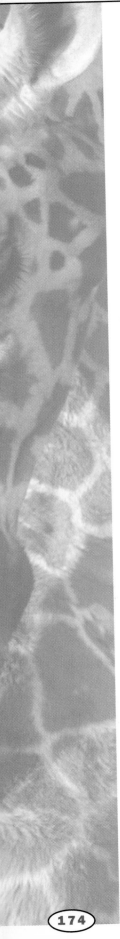

**bilateria:** animals with bilateral symmetry

**bilipid membrane:** a cell membrane that is made up of two layers of lipid or fat molecules

**bio-accumulation:** the build up of toxic chemicals in an organism

**bioactive protein:** a protein that takes part in a biological process

**bioactive proteins:** proteins that take part in biological processes

**biodiversity:** the variety of organisms found in an ecosystem

**biogeography:** the study of the distribution of animals over an area

**biological control:** the introduction of natural enemies such as parastites, predators, or pathogens as a method of controlling pests instead of using chemicals

**biological controls:** introduction of natural enemies such as parastites, predators, or pathogens as a method of controlling pests instead of using chemicals

**biomagnification:** increasing levels of toxic chemicals through each trophic level of a food chain

**biomass:** the dry weight of organic matter comprising a group of organisms in a particular habitat

**biome:** a major type of ecological community

**biometry:** the biological application of statistics to biology

**biotic:** pertaining to living organisms in an environment

**biotic factors:** biological or living aspects of an environment

**bipedal:** walking on two legs

**bipedalism:** describes the ability to walk on two legs

**birthrate:** a ratio of the number of births in an area in a year to the total population of the area

**birthrates:** ratios of the numbers of births in an area in a year to the total population of the area

**bivalve mollusk:** a mollusk with two shells such as a clam

**bivalve mollusks:** mollusks with two shells such as clams

**bivalves:** mollusks that have two shells

**body plan:** the overall organization of an animal's body

**bone tissue:** dense, hardened cells that makes up bones

**botany:** the scientific study of plants

**bovid:** a member of the family bovidae which is hoofed and horned ruminants such as cattle, sheep, goats and buffaloes

**bovids:** members of the family bovidae which are hoofed and horned ruminants such as cattle, sheep, goats and buffaloes

**brachiopods:** a phylum of marine bivalve mollusks

**brackish:** a mix of salt water and fresh water

**brood parasites:** birds who lay their eggs in another bird's nest so that the young will be raised by the other bird

**buccal:** mouth

**budding:** a type of asexual reproduction where the offspring grow off the parent

**buoyancy:** the tendency of a body to float when submerged in a liquid

**Burgess Shale:** a 550 million year old geological formation found in Canada that is known for well preserved fossils

**calcified:** made hard through the deposition of calcium salts

**calcite:** a mineral form of calcium carbonate

**calcium:** a soft, silvery white metal with a chemical symbol of Ca

**capture-recapture method:** a method of estimating populations by capturing a number of individuals, marking them, and then seeing what percentage of newly captured individuals are captured again

**cardiac:** relating to the heart

**cardiac muscle:** type of muscle found in the heart

**cardiopulmonary:** of or relating to the heart and lungs

**carnivorous:** describes animals that eat other animals

**carrying capacity:** the maximum population that can be supported by the resources

**cartilage:** a flexible connective tissue

**cartilaginous:** made of cartilage

**catadromous:** living in freshwater but moving to saltwater to spawn

**character displacement:** a divergence of overlapping characteristics in two species living in the same environment as a result of resource partitioning

**chelicerae:** the biting appendages of arachnids

**chemoreceptors:** a receptor that responds to a specific type of chemical molecule

**chemosynthesis:** obtaining energy and making food from inorganic molecules

**chemosynthetic autotrophs:** an organism that uses carbon dioxide as a carbon source but obtains energy by oxidizing inorganic substances

**chemotrophs:** animals that make energy and produce food by breaking down inorganic molecules

**chitin:** a complex carbohydrate found in the exoskeleton of some animals

**chitinous:** made of a complex carbohydrate called chitin

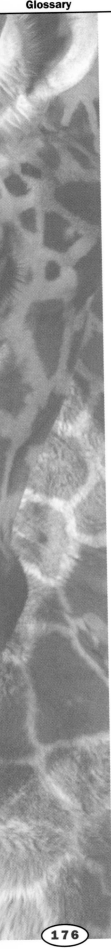

**chloroquine:** a drug commonly used to treat malaria

**chromosomes:** structures in the cell that carry genetic information

**cilia:** hair-like projections used for moving

**circadian rhythm:** daily, 24-hour cycle of behavior in response to internal biological cues

**clades:** a branching diagram that shows evolutionary relationships of organisms

**Class Branchiopoda:** a group of marine bivalve mollusks

**Class Malacostraca:** crustaceans such as lobsters, crabs, and shrimp

**Class Maxillopoda:** crustaceans such as barnacles, ostracods, and copepods

**Class Merostomata:** crustaceans such as horseshoe crabs and eurypterids

**Class Pycnogonida:** crustaceans such as sea spiders

**cleavage:** the process of cytokinesis in animal cells; as cells multiply, the plasma membrane pinches off to make two cells from one

**climate:** long-term weather patterns for a particular region

**cnidaria:** a phylum of aquatic invertebrates such as jellyfishes, corals, sea anemones, and hydras

**cnidarians:** aquatic invertebrates such as jellyfishes, corals, sea anemones, and hydras

**codominance:** an equal expression of two alleles in a heterozygous organism

**codon:** the genetic code for an amino acid that is represented by three nitrogen bases

**codons:** the genetic code for an amino acid that is represented by three nitrogen bases

**coelom:** a body cavity

**coevolution:** a situation in which two or more species evolve in response to each other

**coexist:** live together

**commensal:** a symbiotic relationship wherein which one species benefits and the other is neither helped nor harmed

**competitive exclusion principle:** the concept that when populations of two different species compete for the same limited resources, one species will use the resources more efficiently and have a reproductive edge and eventually eliminate the other species

**compound eye:** a multifaceted eye that is made up of thousands of simple eyes

**compound eyes:** multifaceted eyes that are made up of thousands of simple eyes

**concentric:** having the same center

**conchiolin:** a protein that is the organic basis of mollusk shells

**coniferous, conifers:** having pine trees and other conifers

**connective tissue:** cells that make up bones, blood, ligaments, and tendons

**consumers:** animals that do not make their own food but instead eat other organisms

**continental drift:** the movement of the continents over geologic time

**contour feather:** a feather that covers a bird's body and gives shape to the wings or tail

**contour feathers:** feathers that cover a bird's body and give shape to the wings or tail

**controversy:** a discussion marked by the expression of opposing views

**convergence:** animals that are not closely related but they evolve similar structures

**copulation:** the act of sexual reproduction

**crinoids:** an echinoderm with radial symmetry that resembles a flower

**critical period:** a limited time in which learning can occur

**critical periods:** a limited time in which learning can occur

**crustaceans:** arthropods with hard shells, jointed bodies, and appendages that mainly live in the water

**ctenoid scale:** a scale with projections on the edge like the teeth on a comb

**cumbersome:** awkward

**cytoplasm:** fluid in eukaryotes that surrounds the nucleus and organelles

**cytosolic:** the semifluid portions of the cytoplasm

**death rate:** a ratio of the number of deaths in an area in a year to the total population of the area

**deciduous:** having leaves that fall off at the end of the growing season

**denaturing:** break down into small parts

**dendrites:** branched extensions of a nerve cell that transmit impulses to the cell body

**described:** a detailed description of a species that scientists can refer to identify that species from other similar species

**dessication:** drying out

**detritus:** dead organic matter

**deuterostome:** animal in which the first opening does not form the mouth, but becomes the anus

**deuterostomes:** animals in which the first opening does not form the mouth, but becomes the anus

**diadromous:** animals that migrate between freshwater and saltwater

**differentiation:** differences in structure and function of cells in multicellular organisms as the cells become specialized

**diffusion:** the movement of molecules from a region of higher concentration to a region of lower concentration

**dioecious:** having members of the species that are either male or female

**diploblastic:** having two germ layers; ectoderm and endoderm

**diploid cells:** cells with two sets of chromosomes

**direct fitness:** fitness gained through personal reproduction

**diurnal:** active in the daytime

**DNA replication:** the process by which two strands of a double helix separate and form two identical DNA molecules

**dominance hierarchies:** the structure of the pecking order of a group of individuals of a group where the multiple levels of dominance and submission occur

**dominant:** an allele that is always an expressed trait

**dorsal:** the back surface of an animal with bilateral symmetry

**dorsal root ganglia:** nervous tissue located near the backbone

**dorsoventrally:** flattened from the top and bottom

**dysentery:** inflammation of the intestines that is characterized by pain, diarrhea, and the passage of mucous and blood

**ecdysis:** shedding the outer layer of skin or exoskeleton

**ecdysone:** hormone that triggers molting in arthropods

**echinoderms:** sea animals with radial symmetry such as starfish, sea urchins, and sea cucumbers

**ecological:** relating to an organism's interaction with its environment

**ecology:** study of how organisms interact with their environment

**ecosystem:** a self-sustaining collection of organisms and their environment

**ecosystems:** self-sustaining collections of organisms and their environments

**ecotourism:** tourism that involves travel to areas of ecological or natural interest usually with a naturalist guide

**ectodermal:** relating to the outermost of the three germ layers in animal embryos

**ectoparasite:** an organism that lives on the surface of another organism and derives its nutrients directly from that organism

**ectoparasites:** organisms that live on the surfaces of other organisms and derive their nutrients directly from those organisms

**edentates:** lacking teeth

**El Niño:** a periodic condition characterized by a warming of the central Pacific Ocean and the changes in global weather patterns that are brought about

**emit:** to send out or give off

**endocrine system:** the grouping of organs or glands that secrete hormones into the bloodstream

**endoparasite:** an organism that lives inside another organism and derives its nutrients directly from that organism

**endoparasites:** organisms that live inside other organisms and derive their nutrients directly from those organisms

**endoskeleton:** a skeleton that is surrounded by muscle tissue

**endosymbionts:** the hypothesis that certain organelles in eukaryotes are prokaryotes that have a symbiotic relationship and live within the eukaryote

**endotrophic:** deriving nourishment from within

**enterocoelous:** a cavity formed by the in-folding of the wall of the intestinal cavity in a gastrula

**enzyme:** a protein that acts as a catalyst to start a biochemical reaction

**enzymes:** proteins that act as catalysts to start biochemical reactions

**epidermis:** the protective portion of the outer portion of the skin found in some animals, it is composed of two layers of cells where the outer layer is continuously shed and replaced by the inner layer

**epistasis:** a phenomenon in which one gene alters the expression of another gene that is independently inherited

**epithelial cells:** cells that occur in tightly packed sheets that line organs and body cavities

**epithelial lining:** sheets of tightly packed cells that cover organs and body cavities

**epitope:** a localized region on an antigen that is recognized chemically by antibodies

**equilibrium:** a state of balance

**erythrocytes:** red blood cells, cells containing hemoglobin that carry oxygen throughout the body

**estuaries:** an area of brackish water where a river meets the ocean

**ethology:** animal behavior

**eucoelomates:** animals that have a true body cavity that is completely surrounded by mesoderm

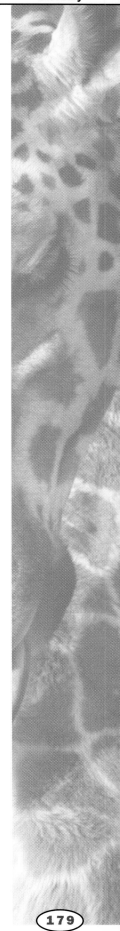

**eukaryota:** a group of organisms containing a membrane bound-nucleus and membrane-bound organelles

**eukaryotes:** organisms containing a membrane-bound nucleus and membrane-bound organelles

**eukaryotic cells:** contains a membrane-bound nucleus and membrane-bound organelles

**euryhaline:** animals that can live in a wide range of salt concentrations

**eusocial:** animals that show a true social organization

**evaporites:** rocks formed from evaporation of salty and mineral-rich liquid

**excrescence:** an abnormal growth

**excrescences:** abnormal growths

**exons:** the coding region in a eukaryotic gene that is expressed

**exoskeleton:** a hard outer protective covering common in invertebrates such as insects

**exoskeletons:** hard outer protective coverings common in invertebrates such as insects

**exponential growth:** a population growing at the fastest possible rate under ideal conditions

**extant:** still living

**facilitated diffusion:** the spontaneous passing of molecules attached to a carrier protein across a membrane

**facultative parasites:** organisms that can survive either as a parasite or free-living

**falconry:** a sport where falcons are used for hunting

**fascicle:** a close cluster

**fauna:** animals

**fertilization:** the fusion of male and female gametes

**fibroblasts:** type of cells found in loose connective tissue that secretes the proteins for connective fibers

**fight or flight response:** an automatic, chemically controlled response to a stimulus that causes increased heart and breathing rates for increased activity

**filter feeders:** animals that strain small food particles out of water

**fission:** dividing into two parts

**fixed action pattern:** behaviors that are common to all members of a species

**flagella (flagellum):** cellular tail that allows the cell to move

**flagellae:** cellular tails that allow cells to move

**flora:** plants

**fossil record:** a collection of all known fossils

**frequency-depentant selection:** a decline in the reproductive success of a particular body type due to that body type becoming common in the population

**frugivores:** fruit-eating animals

**functional morphology:** studying form and function

**fusion:** coming together

**gametes:** reproductive cells that only have one set of chromosomes

**gametocyte:** cell that produces gametes through division

**gametocytes:** cells that produce gametes through division

**ganoid scale:** hard, bony, and enamel covered scales

**gastropods:** mollusks that are commonly known as snails

**gastrovascular cavity:** a single cavity where digestion occurs

**gastrulation:** the formation of a gastrula from a blastula

**gene therapy:** a process where normal genes are inserted into DNA to correct a genetic disorder

**genes:** segments of DNA located on chromosomes that direct protein production

**genetic trait:** trait related to biological inheritance

**genetics:** the branch of biology that studies heredity

**genome:** an organism's genetic material

**genomes:** the sum of all genes in a set of chromosomes

**genotype:** the genetic makeup of an organism

**germ cell:** an egg or sperm cell, a gamete

**germ cells:** egg or sperm cells, gametes

**gill arches:** arches of cartilage that support the gills of fishes and some amphibians

**gill filaments:** the site of gas exchange in aquatic animals such as fish and some amphibians

**gills:** site of gas exchange between the blood of aquatic animals such as fish and the water

**gizzard:** the muscular part of the stomach of some animals where food is ground

**global warming:** a slow and steady increase in the global temperature

**glycoprotein:** an organic molecule that contains a carbohydrate and a protein

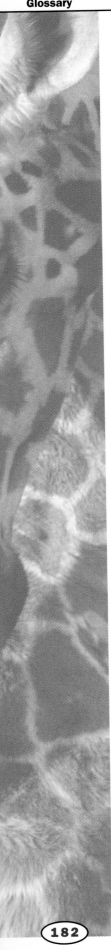

**gonad:** the male and female sex organs that produce sex cells

**gonads:** the male and female sex organs that produce sex cells

**granulocytes:** a type of white blood cell where its cytoplasm contains granules

**green house effect:** a natural phenomenon where atmospheric gases such as carbon dioxide prevent heat from escaping through the atmosphere

**habitat:** the physical location where organisms live in an ecosystem

**habitat loss:** the destruction of habitats through natural or artificial means

**habitat requirement:** necessary conditions or resources needed by an organism in its habitat

**habitats:** physical locations where organisms live in an ecosystem

**Hamilton's Rule:** individuals show less agression to closely related kin than to more distantly related kin

**haplodiploidy:** the sharing of half the chromosomes between a parent and an offspring

**haploid cells:** cells with only one set of chromosomes

**hemocoel:** a cavity between organs in arthropods and mollusks through which blood circulates

**hemocyanin:** respiratory pigment found in some crustaceans, mollusks, and arachnids

**hemoglobin:** an iron-containing protein found in red blood cells that binds with oxygen

**hemolymph:** the body fluid found in invertebrates with open circulatory systems

**herbivore:** an animal that eats plants only

**herbivores:** animals that eat only plants

**herbivorous:** animals that eat plants

**heredity:** the passing on of characteristics from parents to offspring

**heritability:** the ability to pass characteristics from a parent to the offspring

**hermaphodite:** an animals with both male and female sex organs

**hermaphroditic:** having both male and female sex organs

**heterodont:** teeth differentiated for various uses

**heterotrophic eukaryotes:** organisms containing a membrane-bound nucleus and membrane-bound organelles and do not make their own food

**heterotrophs:** organisms that do not make their own food

**heteroxenous:** a life cycle in which more than one host individual is parasitized

**heterozygote:** an organism whose chromosomes contain both genes of a contrasting pair

**heterozygote advantage:** a condition where a heterozygous individual has a reproductive advantage over a homozygous individual

**Hippocrates:** a central figure in medicine in ancient Greece, he is considered the father of modern medicine

**home range:** the area where an animal lives and eats

**homeostasis:** a state of equilibrium in an animal's internal environment that maintains optimum conditions for life

**homeothermic:** describes animals able to maintain their body temperatures

**hominid:** belonging to the family of primates

**hominids:** belonging to the family of primates

**homodont:** teeth with a uniform size and shape

**homologous:** similar but not identical

**homology:** correspondence in the type of structure and its origin

**homoplastic:** similar but of different origins

**homozygote:** an animal with two identical alleles for one trait

**hormone:** a chemical signal secreted by glands that travel through the bloodstream to regulate the body's activities

**hormones:** chemical signals secreted by glands that travel through the bloodstream to regulate the body's activities

**Horseshoe crabs:** "living fossils" in the class of arthropods

**Hox genes:** also known as selector genes because their expression leads embryonic cells through specific morphologic development

**Human Genome Project:** a study by U.S. Department of Energy and the National Institutes of Health to map the entire human genome by 2003

**hunting season:** a period of time during which hunting is permited

**hunting seasons:** periods of time during which hunting is permited

**hybrid:** offspring resulting from the cross of two different species

**hydrostatic skeleton:** a pressurized, fluid-filled skeleton

**hyperpolarizing potential:** any change in membrane potential that makes the inside of the membrane more negatively charged

**hypothalamus:** part of the upper end of the brain stem that regulates activities in the nervous and endrocrine systems

**IgA:** imunoglobin A; a class of proteins that make up antibodies

**IgD:** imunoglobin D; a class of proteins that make up antibodies

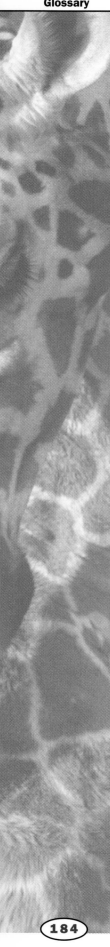

**IgE:** imunoglobin E; a class of proteins that make up antibodies

**IgG:** imunoglobin G; a class of proteins that make up antibodies

**IgM:** imunoglobin M; a class of proteins that make up antibodies

**inbreeding depression:** loss of fitness due to breeding with close relatives

**incomplete dominance:** a type of inheritance where the offspring have an intermediate appearance of a trait from the parents

**incus:** one of three small bones in the inner ear

**indirect fitness:** fitness gained through aiding the survival of non-descendant kin

**infrared:** an invisible part of the electromagnetic spectrum where the wavelengths are shorter than red; heat is carried on infrared waves

**innate behavior:** behavior that develops without influence from the environment

**innervate:** supplied with nerves

**inoculation:** introduction into surroundings that support growth

**insectivore:** an animal that eats insects

**insectivores:** animals that eat insects

**instars:** the particular stage of an insect's or arthropod growth cycle between moltings

**integument:** a natural outer covering

**intercalation:** placing or inserting between

**intraspecific:** involving members of the same species

**introns:** a non-coding sequence of base pairs in a chromosome

**invagination:** a stage in embryonic development where a cell layer buckles inward

**invertebrates:** animals without a backbone

**involuntary muscles:** muscles that are not controlled by will

**isthmus:** a narrow strip of land

**iteroparous:** animals with several or many reproductive events in their lives

***k*-selected species:** a species that natural selection has favored at the carrying capacity

***k*-selecting habitat:** habitat where there is a high cost of reproduction and is sensitive to the size of the offspring

**key innovation:** a modification that permits an individual to exploit a resource in a new way

**keystone species:** a species that controls the environment and thereby determines the other species that can survive in its presence

**krill:** an order of crustaceans that serves as a food source for many fish, whales, and birds

**lancelet:** a type of primitive vertebrate

**lancelets:** primitive vertebrates

**lateral inhibition:** phenomenon that amplifies the differences between light and dark

**lateral line:** a row of pressure sensitive sensory cells in a line on both sides of a fish

**learned behavior:** behavior that develops with influence from the environment

**learning:** modifications to behavior motivated by experience

**leukocytes:** a type of white blood cells that are part of the immune system

**life history strategies:** methods used to overcome pressures for foraging and breeding

**life history strategy:** methods used to overcome pressures for foraging and breeding

**lipids:** fats and oils; organic compounds that are insoluble in water

**logistic growth:** in a population showing exponential growth the individuals are not limited by food or disease

**lungs:** sac-like, spongy organs where gas exchange takes place

**lymphocytes:** white blood cell that completes development in bone marrow

**macroparsite:** a parasite that is large in size

**macroparasites:** parasites that are large in size

**macrophages:** white blood cell that attacks anything foreign such as microbes

**malleus:** the outermost of the inner ear bones

**mantle:** the tissue in mollusks that drapes over the internal organs and may secrete the shell

**mantles:** tissues in mollusks that drape over the internal organs and may secrete the shell

**matrix:** the nonliving component of connective tissue

**megachiroptera:** fruit bats and flying foxes

**meiosis:** a specialized type of cell division that results in four sex cells or gametes that have half the genetic material of the parent cell

**merozoite:** a motile stage in some parastic protozoa

**mesenchyme:** the part of the mesoderm from which the connective tissues (bone, cartilage, and vascular system) arise

**mesenteries:** the membrane that suspends many internal organs in the fluid-filled body cavity of vertebrates

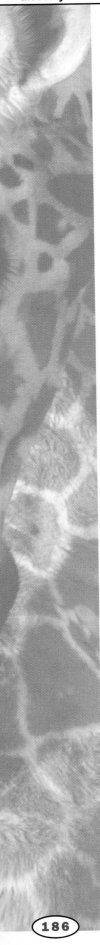

**mesoderm:** the middle layer of cells in embryonic tissue

**messenger RNA:** a type of RNA that carries protein synthesis information from the DNA in to the nucleus to the ribosomes

**metamorphose:** to change drastically from a larva to an adult

**metamorphoses:** changes drastically from its larval form to its adult form

**metamorphosing:** changing drastically from a larva to an adult

**metamorphosis:** a drastic change from a larva to an adult

**metazoan:** a subphylum of animals that have many cells, some of which are organized into tissues

**metazoans:** a subphylum of animals that have many cells, some of which are organized into tissues

**microchiroptera:** small bats that use echolocation

**microparasite:** very small parasite

**microparasites:** very small parasites

**midoceanic ridge:** a long chain of mountains found on the ocean floor where tectonic plates are pulling apart

**mitochondria:** organelles in eukaryotic cells that are the site of energy production for the cell

**Mitochondrial DNA:** DNA found within the mitochondria that control protein development in the mitochondria

**mitosis:** a type of cell division that results in two identical daughter cells from a single parent cell

**modalities:** to conform to a general pattern or belong to a particular group or category

**modality:** to conform to a general pattern or belong to a particular group or category

**molecular clock:** using the rate of mutation in DNA to determine when two genetic groups spilt off

**molecular clocks:** using the rate of mutation in DNA to determine when two genetic groups spilt off

**mollusks:** large phylum of invertebrates that have soft, unsegmented bodies and usually have a hard shell and a muscular foot; examples are clams, oysters, mussels, and octopuses

**molted:** the shedding of an exoskeleton as an animal grows so that a new, large exoskeleton can be secreted

**molting:** the shedding of an exoskeleton as an animal grows so that a new, large exoskeleton can be secreted

**monoculture:** cultivation of a single crop over a large area

**monocultures:** cultivation of single crops over large areas

**monocytes:** the largest type of white blood cell

**monophyletic:** a taxon that derived from a single ancestral species that gave rise to no other species in any other taxa

**monotremes:** egg-laying mammals such as the platypus and echidna

**monoxenous:** a life cycle in which only a single host is used

**morphogenesis:** the development of body shape and organization during ontogeny

**morphological:** the structure and form of an organism at any stage in its life history

**morphological adaptation:** an adaptation in form and function for specific conditions

**morphological adaptations:** adaptations in form and function for specific conditions

**morphologies:** the forms and structures of an animal

**mutation:** an abrupt change in the genes of an organism

**mutations:** abrupt changes in the genes of an organism

**mutualism:** ecological relationship beneficial to all involved organisms

**mutualisms:** ecological relationships beneficial to all involved organisms

**mutualistic relationship:** symbiotic relationship where both organisms benefit

**mutualistic relationships:** symbiotic relationships where both organisms benefit

**mutualists:** a symbiotic relationship where both organisms benefit

**myofibril:** longitudinal bundles of muscle fibers

**myofilament:** any of the ultramicroscopic filaments, made up of actin and myosin, that are the structural units of a myofibril

**myosin:** the most common protein in muscle cells, responsible for the elastic and contractile properties of muscle; it combines with actin to form actomyosin

**natural selection:** the process by which organisms best suited to their environment are most likely to survive and reproduce

**naturalist:** a scientist who studies nature and the relationships among the organisms

**naturalists:** scientists who study nature and the relationships among the organisms

**neuromuscular junction:** the point where a nerve and muscle connect

**neuron:** a nerve cell

**neurons:** nerve cells

**neurotransmitters:** chemical messengers that are released from one nerve cell that cross the synapse and stimulate the next nerve cell

**niche:** how an organism uses the biotic and abiotic resources of its environment

**nocturnal:** active at night

**notochord:** a rod of cartilage that runs down the back of Chordates

**nucleotide:** the building block of a nucleic acid that is composed of a five-carbon sugar bonded to a nitrogen and a phosphate group

**nucleotide chain:** a chain composed of five-carbon sugar groups that forms the basis for nucleic acid

**nucleotides:** building blocks of a nucleic acid that are composed of a five-carbon sugar bonded to a nitrogen and a phosphate group

**obligative mutualism:** an animal that must exist as part of a mutually beneficial relationship

**obligatory parasites:** an animal that can only exist as a parasite

**olfactory:** relates to the sense of smell

**omnivorous:** eating both plants and animals

**ontogeny:** the embryonic development of an organism

**oocyst:** a cyst in sporozoans that contains developing sporozoites

**operculum:** a flap covering an opening

**operculum chamber:** space covered by a flap

**organelles:** membrane-bound structures found within a cell

**ornithology:** the study of birds

**osmoregulatory functions:** controlling the water balance within an animal

**osmoregulatory system:** system that regulates the water balance between an organism and its environment

**osmosis:** the diffusion of water across a membrane

**ossification:** deposition of calcium salts to form hardened tissue such as bone

**osteoblasts:** potential bone forming cells found in cartilage

**oviparous:** having offspring that hatch from eggs external to the body

**ovoviparity:** having offspring that hatch from eggs retained in the mother's uterus

**ovoviviparous:** having offspring that hatch from eggs retained in the mother's uterus

**paleoanthropology:** the study of ancient humans

**parasitology:** the study of parasites

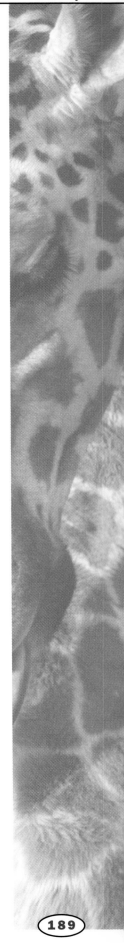

**parasympathetic division:** part of the nervous system that generally enhances body activities that gain and conserve energy such as digestion and heart rate

**parental imprinting:** a process by which a gene's expression in a child depends on which parent donated it before development

**passive diffusion:** the passing of molecules across a membrane from an area of higher concentration to an area of lower concentration without any energy input

**pathogens:** disease-causing agents such as bacteria, fungi, and viruses

**pecking order:** the position of individuals of a group wherein multiple levels of dominance and submission occur

**pectoral:** of, in, or on the chest

**pedipapls:** one pair of short appendages near the mouth in some arthropods used for feeding and copulation

**pericardial cavity:** the space within the membrane that surrounds the heart

**peripheral nervous system:** the sensory and motor nerves that connect to the central nervous system

**peritoneum:** the thin membrane that lines the abdomen and covers the organs in it

**pesticide:** any substance that controls the spread of harmful or destructive organisms

**pesticides:** substances that control the spread of harmful or destructive organisms

**pH:** a measure of how acidic or basic a substance is by measuring the concentration of hydrogen ions

**phalanges:** bones of the fingers and toes

**pharyngeal:** having to do with the tube that connects the stomach and the esophagus

**phenotype:** physical and physiological traits of an animal

**phenotypes:** the physical and physiological traits of an animal

**phenotypic:** describes the physical and physiological traits of an animal

**phenotypic trait:** physical and physiological variations within a population

**phenotypic variation:** differences in physical and physiological traits within a population

**pheromones:** small, volatile chemicals that act as signals between animals that influence physiology or behavior

**phlogenetic:** relating to the evolutionary history

**phospholipid:** molecules that make up double layer membranes; one end of the molecule attracts water while the other end repels water

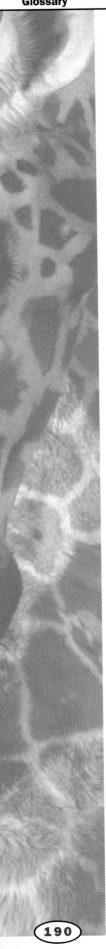

**photoreceptors:** specialized cells that detect the presence or absence of light

**photosynthesis:** the combination of chemical compounds in the presence of sunlight

**photosynthesizing autotrophs:** animals that produce their own food by converting sunlight to food

**phyla:** broad, principle divisions of a kingdom

**phylogenetic:** relating to the evolutionary history of species or group of related species

**phylogeny:** the evolutionary history of a species or group of related species

**physiological:** relating to the basic activities that occur in the cells and tissues of an animal

**physiology:** the study of the normal function of living things or their parts

**placenta:** the structure through which a fetus obtains nutrients and oxygen from its mother while in the uterus

**placental:** having a structure through which a fetus obtains nutrients and oxygen from its mother while in the uterus

**placoid scale:** a scale composed of three layers and a pulp cavity

**placoid scales:** scales composed of three layers and a pulp cavity

**plankton:** microscopic organisms that float or swim weakly near the surface of ponds, lakes, and oceans

**plate tectonics:** the theory that Earth's surface is divided into plates that move

**platelet:** cell fragment in plasma that aids clotting

**platelets:** cell fragments in plasma that aid in clotting

**pleural cavity:** the space where the lungs are found

**plumose:** having feathers

**pluripotent:** a cell in bone marrow that gives rise to any other type of cell

**poaching:** hunting game outside of hunting season or by using illegal means

**poikilothermic:** an animal that cannot regulate its internal temperature; also called cold blooded

**polymer:** a compound made up of many identical smaller compounds linked together

**polymerase:** an enzyme that links together nucleotides to form nucleic acid

**polymerases:** enzymes that link together nucleotides to form nucleic acid

**polymodal:** having many different modes or ways

**polymorphic:** referring to a population with two or more distinct forms present

**polymorphism:** having two or more distinct forms in the same population

**polymorphisms:** having two or more distinct forms in the same population

**polyploid:** having three or more sets of chromosomes

**polysaccharide:** a class of carbohydrates that break down into two or more single sugars

**polysaccharides:** carbohydrates that break down into two or more single sugars

**population:** a group of individuals of one species that live in the same geographic area

**population density:** the number of individuals of one species that live in a given area

**population dynamics:** changes in a population brought about by changes in resources or other factors

**population parameters:** a quantity that is constant for a particular distribution of a population but varies for the other distrubutions

**populations:** groups of individuals of one species that live in the same geographic area

**posterior:** behind or the back

**precursor:** a substance that gives rise to a useful substance

**prehensile:** adapted for siezing, grasping, or holding on

**primer:** short preexisting polynucleotide chain to which new deoxyribonucleotides can be added by DNA polymerase

**producers:** organisms which make up the level of an ecosystem that all other organisms ultimately depend on; usually these are plants

**progeny:** offspring

**prokaryota:** a group of organisms that lack a membrane-bound nucleus and membrane-bound organelles

**prokaryotes:** single-celled organisms that lack a true cell nucleus

**prokaryotic endosymbionts:** single-celled organisms that lack a true cell nucleus that live inside of other cells

**proprioceptors:** sense organs that receive signals from within the body

**protostome:** animal in which the initial depression that starts during gastrulation becomes the mouth

**protostomes:** animals in which the initial depression that starts during gastrulation becomes the mouth

**protozoa:** a phylum of single-celled eukaryotes

**protozoan:** a member of the phylum of single-celled organisms

**pseudocoelom:** a body cavity that is not entirely surrounded by mesoderm

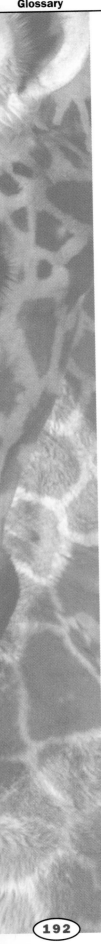

**pseudocoelomates:** animals with a body cavity that is not entirely surrounded by mesoderm

**pterylae:** feather tracks

**quadrupedal:** describes an animal with four legs

**quadrupeds:** animals with four legs

**quinine:** substance used to treat malaria

***r*-selected species:** a species that shows the following characteristics: short lifespan; early reproduction; low biomass; and the potential to produce large numbers of usually small offspring in a short period of time

***r*-selecting habitat:** the concept where a high reproductive rate is the chief determinant of life history

**radially symmetric:** wheel-like symmetry in which body parts radiate out from a central point

**radially symmetrical:** describes an animal that features a wheel-like symmetry in which its body parts radiate out from a central point

**recessive:** a hidden trait that is masked by a dominant trait

**recombinant DNA:** DNA that is formed when a fragment of DNA is incorporated into the DNA of a plasmid or virus

**regeneration:** regrowing body parts that are lost due to injury

**relative abundance:** an estimate of population over an area

**rennin:** an enzyme used in coagulating cheese; is obtained from milk-fed calves

**resorbed:** absorption of materials already in the body

**resorption:** absorbing materials that are already in the body

**respiratory pigments:** any of the various proteins that carry oxygen

**restriction enzymes:** bacterial proteins that cut DNA at specific points in the nucleotide sequence

**retina:** a layer of rods and cones that line the inner surface of the eye

**riparian:** habitats in rivers and streams

**ruminants:** plant-eating animals with a multicompartment stomach such as cows and sheep

**sagital plane:** a plane that runs long-ways through the body

**salamanders:** four-legged amphibians with elongated bodies

**sarcomere:** one of the segments into which a fibril of striated muscle is divided by thin dark bands

**scavengers:** animals that feed on the remains of animals it did not kill

**schizocoelous:** the mesoderm originates from existing cell layers when the cells migrate

**scleroblasts:** cells that give rise to mineralized connective tissue

**sedimentary rock:** rock that forms when sediments are compacted and cemented together

**semelparous:** animals that only breed once and then die

**serial homology:** a rhythmic repetition

**sessile:** not mobile, attached

**sexual reproduction:** a reproduction method where two parents give rise to an offspring with a different genetic makeup from either parent

**sexual selection:** selection based on secondary sex characteristics that leads to greater sexual dimorphorphism or differences between the sexes

**sexual size dimorphism:** a noticeable difference in size between the sexes

**shoals:** shallow waters

**single-lens eyes:** an eye that has a single lens for focusing the image

**skeletal muscle:** muscle attached to the bones and responsible for movement

**smooth muscle:** muscles of internal organs which is not under conscious control

**somatic:** having to do with the body

**somatic nervous system:** part of the nervous system that controls the voluntary movement of skeletal muscles

**somatosensory information:** sensory information from different parts of the body except for the eyesm tongue, ears, and other primary sense organs

**somites:** a block of mesoderm along each side of a chordate embryo

**sonar:** the bouncing of sound off distant objects as a method of navigation or finding food

**spinal cord:** thick, whitish bundle of nerve tissue that extends from the base of the brain to the body

**splicing:** spliting

**spongocoel:** the central cavity in a sponge

**sporozoa:** a group of parasitic protozoa

**sporozoans:** parasitic protozoans

**sporozoite:** an infective stage in the life cycle of sporozoans

**stapes:** innermost of the three bones found in the inner ear

**stimuli:** anything that excites the body or part of the body to produce a specific response

**stimulus:** anything that excites the body or part of the body to produce a specific response

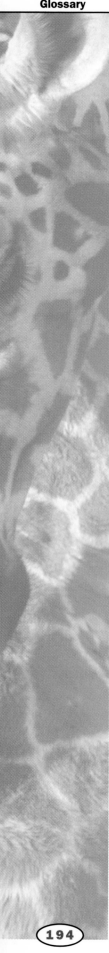

**strata:** layers of sedimentary rock consisting of approximately the same kinds of material

**striated muscle:** a type of muscle with fibers of cross bands usually contracted by voluntary action

**striated muscles:** muscles with fibers of cross bands usually contracted by voluntary actions

**superposition:** the order in which sedimentary layers are found with the youngest being on top

**symbiosis:** any prolonged association or living together of two or more organisms of different species

**symbiotic relationship:** close, long-term relationship where two species live together in direct contact

**symbiotic relationships:** close, long-term relationships where two species live together in direct contact

**symmetrical:** a balance in body proportions

**synapse:** the space between nerve cells across which impulses are chemically transmitted

**systematic:** study of the diversity of life

**tactile:** the sense of touch

**tapetum:** a reflective layer in the eye of nocturnal animals

**taxa:** named taxonomic units at any given level

**taxon:** named taxonomic unit at any given level

**taxonomy:** the science of classifying living organisms

**terraria:** a small enclosure or closed container in which selected living plants and sometimes small land animals, such as turtles and lizards, are kept and observed

**terrariums:** small enclosures or closed containers in which selected living plants and sometimes small land animals, such as turtles and lizards, are kept and observed

**terrestrial:** living on land

**thoracic:** the chest area

**thromboplastin:** a protein found in blood and tissues that promotes the conversion of prothrombin to thrombin

**torpid:** a hibernation strategy where the body temperature drops in relation to the external temperature

**trachea:** the tube in air-breathing vertebrates that extends from the larynx to the bronchi

**transcription:** process where enzymes are used to make an RNA copy of a strand of DNA

**transgenic:** an organism that contains genes from another species

**transgenic organism:** an organism that contains genes from another species

**translation:** process where the order of bases in messenger RNA codes for the order of amino acids in a protein

**transverse plane:** a plane perpendicular to the body

**trilobites:** an extinct class of arthropods

**triploblasts:** having three germ layers; ectoderm, mesoderm, and endoderm

**trophic level:** the division of species in an ecosystem by their main source of nutrition

**trophic levels:** divisions of species in an ecosystem by their main source of nutrition

**ungulates:** animals with hooves

**urea:** soluble form of nitrogenous waste excreted by many different types of animals

**urethra:** a tube that releases urine from the body

**uric acid:** insoluble form of nitrogenous waste excreted by many different types of animals

**ventral:** the belly surface of an animal with bilateral symmetry

**vertebrates:** animals with a backbone

**viviparity:** having young born alive after being nourished by a placenta between the mother and offspring

**viviparous:** having young born alive after being nourished by a placenta between the mother and offspring

**vocalization:** the sounds used for communications

**voluntary muscles:** a type of muscle with fibers of cross bands usually contracted by voluntary action

**wavelength:** distance between the peaks or crests of waves

**zooplankton:** small animals who float or weakly move through the water

**zygote:** a fertilized egg

**zygotes:** fertilized eggs

**zymogens:** inactive building-block of an enzyme

# Topic Outline

## ADAPTATIONS

Adaptation
Antlers and Horns
Aposematism
Biological Evolution
Biomechanics
Blood
Camouflage
Catadromous—Diadromous and Anadromous Fishes
Colonization
Communication
Community Ecology
Comparative Biology
Defense
Echolocation
Egg
Extremophile
Locomotion
Mimicry
Peppered Moth
Tool Use
Water Economy in Desert Organisms

## AGRICULTURE

Apiculture
Aquaculture
Classification Systems
Dinosaurs
Domestic Animals
Farmer
Farming
Selective Breeding
Sustainable Agriculture

## ANIMAL DIVERSITY

Animal
Biodiversity
Biogeography
Biological Evolution
Cambrian Explosion
Camouflage
Cephalization
Coevolution
Colonization
Community Ecology
Constraints on Animal Development
Diversity of Major Groups
Extremophile
Functional Morphology
Kingdoms of Life
Phylogenetic Relationships of Major Groups
Phylogenetics Systematics
Prokaryota
Sexual Dimorphism
Taxonomy

## ANIMAL GROUPS

Agnatha
Amphibia
Annelida
Arthropoda
Aves
Cephalochordata
Cestoda
Chondricthyes
Chordata
Cnidaria
Dinosaurs
Echinodermata

Blood
Cancer
Cell Division
Cells
Digestion
Egg
Homeostasis
Hormones
Keratin
Molecular Biologist
Molecular Biology
Molecular Systematics
Physiologist
Physiology
Respiration
Transport

## BIODIVERSITY

Biodiversity
Biogeography
Biomass
Biomes
Colonization
Community Ecology
Diversity of Major Groups
Eukaryota
Habitat
Habitat Loss
Habitat Restoration
Zooplankton

## CAREERS IN ANIMAL SCIENCE

Ecologist
Environmental Lawyer
Farmer
Functional Morphologist
Geneticist
Horse Trainer
Human Evolution
Livestock Manager
Marine Biologist
Medical Doctor
Molecular Biologist
Museum Curator
Paleontologist
Physiologist
Scientific Illustrator

Service Animal Trainer
Systematist
Taxonomist
Veterinarian
Wild Game Manager
Wildlife Biologist
Wildlife Photographer
Zoologist

## CELL BIOLOGY

Absorption
Blood
Cell Division
Cells
Viruses

## ECOLOGY

African Cichlid Fishes
Behavioral Ecology
Biotic Factors
Camouflage
Community Ecology
Competition
Competitive Exclusion
Conservation Biology
DDT
Ecologist
Ecology
Ecosystem
Evolutionary Stable Strategy
Exotic Species
Expenditure per Progeny
Feeding Strategies
Fitness
Food Web
Foraging Strategies
Growth And Differentiation of the Nervous System
Habitat
Habitat Loss
Habitat Restoration
Home Range
Human Commensals and Mutual Organisms
Interspecies Interactions
Iteroparity and Semelparity
Keystone Species
Life History Strategies

Malthus, Thomas Robert
Parasitism
Plankton
Population Dynamics
Populations
Predation
Territoriality
Trophic Level
Zooplankton

## ENVIRONMENT

Biological Pest Control
Biomass
Biomes
Biotic Factors
Carson, Rachel
DDT
Ecosystem
Endangered Species
Environment
Environmental Degradation
Environmental Impact
Environmental Lawyer
Fossil Fuels
Global Warming
Human Populations
Natural Resources
Pesticide
Pollution
Silent Spring
Threatened Species

## ETHICS

Animal Rights
Animal Testing
Bioethics

## EVOLUTION

Adaptation
African Cichlid Fishes
Aposematism
Biological Evolution
Camouflage
Coevolution
Constraints on Animal Development
Continental Drift
Convergence

Darwin, Charles
Genetic Variation in a Population
Heterochrony
Homology
Human Evolution
Lamarck
Leakey, Louis and Mary
Modern Synthesis
Morphological Evolution in Whales
Morphology
Natural Selection
Peppered Moth
Sexual Dimorphism
Sexual Selection
Spontaneous Generation

## FORM AND FUNCTION

Acoustic Signals
Adaptation
African Cichlid Fishes
Antlers and Horns
Aposematism
Biomechanics
Blood
Body Cavities
Body Plan
Bone
Burgess Shale and Ediacaran Faunas
Camouflage
Cell Division
Cells
Cephalization
Chitin
Circulatory System
Communication
Defense
Digestion
Digestive System
Echolocation
Endocrine System
Excretory and Reproductive Systems
Feeding
Flight
Gills
Gliding and Parachuting
Locomotion
Mimicry

Nervous System
Respiratory System
Sexual Selection
Shells
Vision
Vocalization

## GENETICS

Drosophila
Genes
Genetic Engineering
Genetic Variation in a Population
Genetically Engineered Foods
Geneticist
Genetics
Mendel, Gregor
Modern Synthesis
PCR
Viruses

## GEOLOGIC HISTORY

Cambrian Period
Carboniferous
Continental Drift
Cretaceous
Devonian
Geological Time Scale
Jurassic
K/T Boundary
Oligocene
Ordovician
Permian
Pleistocene
Quaternary
Silurian
Tertiary
Triassic

## GROWTH AND DEVELOPMENT

Allometry
Antlers and Horns
Body Cavities
Body Plan
Bone
Cartilage
Cell Division
Cells

Comparative Biology
Constraints on Animal Development
Egg
Embryology
Embryonic Development
Haeckel's Law of Recapitulation
Heterochrony
Mesenchyme
Metamorphosis
Molting
Ontogeny
Serial Homology
Von Baer's Law

## HISTORICAL FIGURES IN SCIENCE

Aristotle
Bailey, Florence Augusta Merriam
Bates, Henry Walter
Carson, Rachel
Darwin, Charles
Diamond, Jared
Elton, Charles Sutherland
Fausto–Sterling, Anne
Fossey, Dian
Goodall, Jane
Gould, Steven Jay
Haldane, J. B. S.
Lamarck, Jean-Baptiste
Leakey, Louis and Mary
Linnaeus, Carolus
Lorenz, Konrad
Malthus, Thomas Robert
Mayr, Ernst
McArthur, Robert
Mendel, Gregor
Montalcini, Rita Levi
Pasteur, Louis
Simpson, George Gaylord
Stevens, Nettie Maria
Wallace, Alfred Russel
Wilson, E. O.

## HUMANS AND THE ANIMAL WORLD

Cultures and Animals
Human Commensals and Mutual Organisms
Human Populations
Human–Animal Conflicts

Hunter-Gatherers
Hunting
Malaria

## LIFE CYCLES

Catadromous—Diadromous and Anadromous
Fishes
Cell Division
Colonization
Courtship
Endosymbiosis
Iteroparity and Semelparity
Malaria
Metamorphosis
Parasitism

## REPRODUCTION

Antlers and Horns
Asexual And Sexual Reproduction
Cell Division

Excretory and Reproductive Systems
Fertilization

## SCIENTIFIC FIELDS OF STUDY

Behavioral Ecology
Community Ecology
Comparative Biology
Conservation Biology
Ecology
Embryology
Entomology
Functional Morphology
Herpetology
Icthyology
Molecular Biology
Morphology
Mouth, Pharynx, and Teeth
Paleontology
Physiology
Sociobiology
Taxonomy

# Cumulative Index

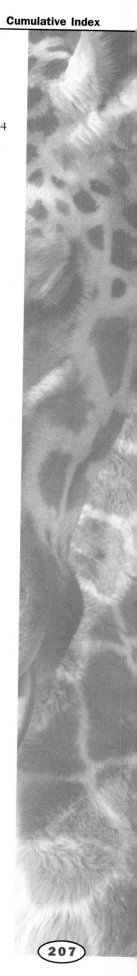

ASPCA (American Society for the Prevention of Cruelty to Animals), 1:24, 1:26, 1:27

Aspidobotharia, 4:123

Assortative mating, 1:76

Asteroidea (starfish and sea stars)
  body plan, 2:36
  as keystone species, 2:37, 2:129, 3:64–65
  See also Echinodermata

Asteroids, role in extinctions, 2:3, 2:25, 3:61–62

Astrocytes, function, 3:153

Asymmetrical body plans, 1:107

Asymmetrical brain development, 3:36

Asymmetrical gaits, 3:85

Atacama Desert, 1:89

Atheriniformes, 3:170

Athletic competition
  biomechanics, 1:84
  and steroids, 2:67

Atmosphere
  biogeochemical cycles, 2:46, 2:48
  pollution (see Air pollution)

ATP (adenosine triphosphate)
  for active transport, 4:120
  defined, 1:1
  metabolism, 2:84
  in mitochondria, 1:136, 2:84
  oxygen and, 4:52
  phosphorous, 2:48

Atriopore, tunicates, 4:130

Atrium, 1:152

Auditory signals. See Acoustic signals

Audubon, John James, 4:75

Audubon Society
  Aldo Leopold and, 3:4
  Florence Bailey and, 1:57
  habitat restoration, 3:1
  on protecting endangered animals, 4:159

Augmentation of natural enemies, 1:79–80

Aural. See Hearing

Australia
  Aborigines, hunting practices, 3:47
  continental drift, 1:73–74, 1:181, 3:163, 4:108
  early arthropods, 4:84
  glacial scouring, 4:1
  marsupials, 1:182, 1:183, 3:102, 4:108

Australopithecines, 3:34–35, 3:38

*Australopithecus*, 4:39
  *aethiopicus*, 3:35

*afarensis* (*garbi*), 3:35, 3:38

*africanus*, 3:35, 3:38

*anamensis*, 3:35

*boisei*, 3:35, 3:38, 3:71

*robustus*, 3:35, 3:38

Autecology, defined, 2:43

Automobile safety, biomechanics, 1:85

Automobiles, and air pollution, 4:19, 4:113

Autonomic nervous system, 3:143, 3:155–156

Autotrophs (producers)
  algae, 2:46, 4:12, 4:128–129
  chemosynthetic, 2:110, 2:127
  defined, 1:95
  metabolism, 4:11
  photosynthesizing, 2:46, 2:110, 2:112, 2:126–128, 4:126–127
  producers, defined, 2:112
  prokaryotes as, 4:42
  role in food web, 1:167, 2:126–127, 4:126–127
  See also Plants

Avery, Oswald, 2:142

Aves (birds), 1:52–56
  adaptations to deserts, 4:148
  classification, 1:52–53
  cooling systems, 1:54, 1:55
  cursorial, 3:86
  disease transmission by, 4:137
  diving locomotion, 3:90
  echolocation, 2:38
  eggs, 1:56, 1:187–188, 2:48–50, 2:53, 2:63
  energy requirements, 2:123–124
  Ernst Mayr's studies, 3:106
  exotic species, 2:92, 2:94
  falconry, 3:46
  fertilization, 2:88
  flight (see Flight, birds)
  homeostasis, 3:14
  homeothermism, 1:52–53
  life history strategies, 3:93
  maritime, role in food web, 2:127
  mistletoe transmission, 3:177
  number of species, 2:27
  oil spills, 4:21, 4:23
  overpredation, 2:61–62, 3:48
  pesticides and, 2:11–12, 2:41, 4:4–5, 4:82–83
  as pets, 2:5, 4:113
  as predators, 1:158–159, 3:163, 3:186–187
  vocalization, 4:142
  waste products, 2:89, 4:149

See also Feathers; *specific birds and bird groups*

Aves, behavior
  acoustic signals, 1:4, 1:5, 1:62, 1:165, 1:186, 1:188
  aggression, 1:12–13
  brood parasitism, 3:51, 3:55, 3:177–178
  care of young, 1:5, 2:22, 2:25
  character displacement, 1:176
  circadian rhythm, 1:149
  commensalism, 1:170
  competition, 1:172, 1:175–176, 3:93
  courtship (see Courtship, birds)
  dominance hierarchies, 2:31
  egg-retrieval instinct, 3:52
  imprinting, 1:60–61, 1:63, 1:189, 2:83, 3:49–51
  instinct, 3:53
  learning, 1:8, 1:56
  migration, 2:184, 3:117, 3:118, 3:119
  mutualistic relationships, 3:54
  navigation, 3:120
  niche partitioning, 1:174
  pair bonding, 1:186–188, 4:76–80
  territoriality, 1:4, 1:170
  tool use, 3:73

Aves, evolution
  adaptations, bone, 1:54, 1:184–185, 4:54, 4:87
  adaptive radiation, 1:8, 1:9
  dinosaur ancestors, 2:25, 3:60, 4:62
  eggs, 1:52
  phylogenetic tree, 4:6, 4:7, 4:105–106
  Pleistocene, 4:16
  posture, 2:24
  synapomorphies and apomorphies, 4:8
  Tertiary, 4:108

Aves, morphology
  air sacs, 1:54, 4:54, 4:142
  bills, 3:140
  circulatory systems, 1:152
  digestive systems, 2:22
  eyes, 4:140
  hox genes, 4:72
  respiration, 4:54, 4:57
  scales, 4:62
  sense organs, 1:55–56, 1:165
  sexual dimorphism, 4:74–75
  teeth, lack of, 3:100, 4:133
  vision, 4:140

Axelrod, Robert, 2:87

interspecific interactions, 1:169, 2:44

Modern Synthesis, 3:105–106, 3:123–125

modifications to theory, 1:75, 2:172

natural selection, 2:11, 3:6, 3:98, 3:149

peppered moth as evidence, 3:185–186

role of mutations, 1:45

sexual selection, 3:150

use of fossil records, 2:136

*Dasyatididae* (stingrays), 1:147

Dawkins, Richard, 2:86

DDT, 2:**11–13**

aerial spraying, *2:12*

banning, 2:13, 4:5

biomagnification, 2:11–12, 4:4–5

as cause of population decline, 2:63

and Dutch elm disease, 2:12

effect on birds, 2:11–12, 4:4–5

effect on fish, 2:12–13

impact on environment, 1:78

invention of, 4:3–4

and malaria, 2:13, 4:4

as neurotoxin, 2:11

Rachael Carson's warnings, 1:128, 2:13, 4:4–5, 4:82–83

resistance to, 4:4

Death rates

defined, 3:75

iteroparity and semelparity, 3:57

and population dynamics, 3:75, 4:25–26

Type III survivorship patterns, 3:75–76

Decapoda. *See* Crabs

Deciduous, defined, 1:90

Decomposers

bacteria, 4:20, 4:22

detrivores, 1:96, 2:112, 2:128

Devonian, 2:17

insects, 1:51, 2:71, 4:93

material cycling, 2:46–48

methane production, 2:133

nematodes, 3:150

role in food web, 2:128

Dedifferentiation, 4:157–158

*Deep Flight I*, 1:95

Deer

antlers, 1:35, 3:9, 4:74

crepuscular, 1:189

desert, 1:89

fossil record, 4:109

Key, as threatened, 4:113

population control, 2:45, 3:26, 3:45–47, 3:66

as ruminants, 2:69

sexual dimorphism, 4:74, 4:76

Defenders of Wildlife Compensation Trust, 3:26, 3:27

Defense, 2:**13–16**, *2:14*

anti-capture methods, 2:15

antlers and horns, 1:36

aposematic signals, 1:38–41, 1:164

cephalopods, 3:133

chemical, 2:14

electrical, 1:147

escape, 1:122

eusocial animals, 4:92–94

evolution in Ordovician, 3:166

gastropods, 3:131

hiding behaviors, 2:13–14

by host plants, 3:129

as instinctive, 3:51, 3:53

nematocysts, 1:108–109, 1:154–155

regenerated body parts, 2:36, 2:37

*schreckstoff*, 3:170

spines and quills, *2:14*, 2:36, 3:63, 3:121–122

*Stegosaurus*, 3:60

surviving attack, 2:15–16

teeth and tusks, 3:141

of territory (*see* Territoriality)

*See also* Acoustic signals; Camouflage; Mimicry; Toxic substances, naturally occurring

Deforestation

environmental impact, 2:73–74, 2:80, 2:181–182, 2:184

and extinctions, 2:61, 2:63, 2:64, 4:113

impact on carbon cycle, 2:73–74

from logging (*see* Logging)

statistics, 1:177

tropical rain forests, 2:80, 2:181, 4:100

Degree of relatedness, defined, 2:119

Delayed sleep phase syndrome (DSPS), 1:150

Demeter, mythological character, 2:105

*Demodex*, as human commensal, 3:30

Demospongiae, 4:35

Denali National Park, 1:*88*

Denature (DNA), defined, 3:182

Dendrites, 2:173, 2:175, 3:154, 3:159, 3:*160*, 4:67

*Dendrobates* (arrow-poison frogs), 2:14

Dendrochronology, as indicator of climate, 4:16

Dengue fever, transmission, 2:26, 2:72

Denitrification, 2:48

Density-dependent growth. *See* Logistic growth model

Density-dependent populations, 4:33

Density-independent populations, 4:33

Dental caries, 3:30, 3:140

Dentin, origin, 3:110

Deoxyribonucleic acid. *See* DNA

Dermal denticles. *See* Placoid scales

*The Descent of Man*, 3:32

Desertification

causes, 2:74

environmental impact, 2:74, 2:79

global issues, 2:77

Pleistocene, 4:44

Deserts

adaptations, 1:6–7, 1:22, 1:89, 4:146–149

Arches National Park, 1:*89*

Atacama, 1:89

biome, 1:89

biome, map, 1:*86*

crepuscular species, 1:189–190

Pangaea, Permian, 4:1–2

productivity, 2:46

recreated in zoological parks, 4:160

Desiccation (dehydration)

of eggs, 2:49

exoskeleton to prevent, 1:143, 4:87

and hydrostatic skeletons, 4:88

and respiratory surfaces, 4:55

scales to prevent, 4:61

Detritus

defined, 2:112

role in food web, 1:96, 2:128

Detrivores, 1:96, 2:112, 2:128, 4:12

Deuterostomes

body plans, 1:106, 3:115

coelom development, 1:101–103

defined, 2:35

echinoderms as, 2:35, 2:56

embryonic development, 1:107, 2:53, *2:54*, 2:56–57, *2:59*

Devonian period, 2:**16–17**, 2:162

adaptive radiation, 2:16–17

Age of Fishes, 1:145, 2:16

cooperative communities, 2:17

invasion of terrestrial vertebrates, 1:19

mass extinctions, 2:99

surrounding time periods, 2:*16*

biogeochemical cycles, 2:46–48

biotic and abiotic factors, 1:85, 1:95–96

carnivore roles, 2:111

climax communities, 3:76

coral reefs, 1:156, 1:167, 2:47

defined, 2:41

earthworm role, 1:32

energy flow, 2:45–46, 2:126–128

equilibrium, 3:64

fire-dependent, 2:183

grasslands, 2:112, 2:128, 4:127–128

insect role, 1:51

loss of, 2:75, 2:80

nitrogen cycles, 2:47–48

population growth, impact, 3:24–25

restoring, 3:1–4

role of extinctions, 2:101

role of plankton, 4:12–13

water cycles, 2:47

zoological parks, recreation, 4:160

See also Biomes; Carrying capacity; Food web; Habitat

Ecotourism, 2:184, 3:27–28, 3:28, 4:151

Ectoderm

defined, 2:55

embryonic development, 1:103, 1:138, 2:52–53, 2:55, 2:58–60, 2:173

Ectoparasites

characteristics, 3:175–176

defined, 1:32

leeches, 1:32

Momogea, 4:13–14

ticks and mites, 1:38, 1:47, 1:48, 3:175

Edentates, evolution, 3:164

The Edge of the Sea, 1:128, 4:82

Ediacaran faunas, 1:115

Eelgrass, diseases of, 2:63

Eels (Anguillidae), 3:168–169

locomotion, 3:83, 3:90

migration, 3:117–118, 3:169

moray, 3:169

Egg-cleavage patterns, 1:107, 2:52–53, 2:56–58, 2:60

Eggs, 2:48–50, 2:50

albumen, 2:50, 2:53, 2:58

allantois, 2:49–50, 2:53, 2:60

amnion, 1:52, 2:49–50, 2:53, 4:112

brood parasitism, 3:51, 3:55, 3:177–178

defined, 2:48

DNA, 2:53, 2:57, 2:88, 2:139, 4:46–47

evolution, 1:52

implantation, 2:60

isolecithal, 2:53

oviparous, 1:144, 2:49, 3:101, 4:51

ovoviparous, 1:144, 3:113

produced by gonads, 2:117

protecting and incubating, 1:187–188, 3:171, 4:75, 4:80

retrieval instinct, 3:52

viviparous, 1:22, 1:144, 2:49, 3:113, 4:51

waste products, 2:49–50, 2:53, 2:89

yolk, 2:49–50, 2:53, 2:57, 2:58, 2:60

See also Embryonic development; Expenditure per progeny; Fertilization

Eggs, specific animal groups

amphibians, 1:19, 1:126–127, 2:53, 4:112

arthropods, 2:53

birds, 1:52, 1:56, 1:187–188, 2:48–50, 2:53, 2:63

chickens, 2:58

Chondricthyes, 2:53

crocodilians, 4:50

dinosaurs, 2:25

Drosophila, 2:33–34

frogs, 2:57

insects, 2:71

invertebrates, 2:53

killifishes, 3:171

mammals, 2:53, 3:100–101

salamanders, 1:21

snakes, 4:51

tapeworms, 1:142

tuataras, 4:50

turtles, 4:49

Eggshells, 2:49–50

effect of DDT, 2:12, 2:63, 4:4

embryonic development, 2:49–50, 2:53, 2:58

lizards, 4:51

Egrets, cattle, 2:90

Egypt (ancient)

animals in folklore, 2:105

ichthyology studies, 3:48

reverence for cats, 2:5, 2:30

Eight Little Piggies, 2:172

El Niño, and malaria, 3:97

Eldredge, Niles, 1:77

Electric currents, communication, 1:166

Electric eels, 1:166

Electric fields, ability to sense, 4:67

Electric fishes (Marmyridae), 3:158, 3:168

Electrocytes, 1:166

Electrophoresis, as research tool, 3:127

Elephantiasis, 3:152, 3:177

Elephants

Asian, toolmaking, 4:114

competition for water, 1:173

cooling systems, 1:173, 2:138

as entertainment, 2:7

expenditure per progeny, 2:97

feeding in zoos, 4:160

feeding strategies, 2:115

functional morphology, 2:138

Indian, as endangered, 3:48

life history strategies, 3:76, 3:93

as modern tetrapods, 4:112

tusks, 3:141

Elevation, role in biodiversity, 1:67

Elk

American, sexual dimorphism, 4:75

functional morphology, 2:138

population control, 3:26

teeth, 3:141

territoriality, 4:105

Elm trees, Dutch elm disease, 2:12

Elopomorpha, 3:168–169

Elton, Charles Sutherland, 2:51–52

Embryology, 2:52

von Baer's law, 4:143–144

Xenopus as test animal, 2:52, 2:57–58, 4:155–157

Embryonic development, 2:53–61, 2:54, 2:59

anterior/posterior polarity, 1:107, 2:34, 2:54, 2:57, 2:58, 2:175

apomorphies, 1:184

bones, 1:111

cartilage, 1:129

circulatory systems, 1:152

coelom development, 1:102–103

convergent extension, 2:54, 2:55, 2:58

endotrophic larvae, 3:111

exotrophic larvae, 3:111

gastraea, postulated, 3:6

gene therapy and screening, 2:151, 2:158

gills, 4:133, 4:144

heterochrony, 3:8–10

intercalation, 2:54–55

invagination, 2:54, 2:57–58

mesenchymal cells, 3:110

mutations, 1:178–179

internal fertilization, 2:88

iteroparous, 3:58

placenta, 1:183, 2:49, 3:100–104

sperm, 2:118

umbilical cord, 2:60

Mammary glands, 3:100, 3:*103*

Mammoths, 4:15, 4:34, 4:44

*Man Meets Dog*, 3:92

Manatees, as threatened, 4:113

Mandibles, insects, 3:141

Manic depression, 1:28

Manta rays (*Mobulidae*), 1:*145*, 1:147

Mantle

mollusks, 1:*108*, 1:109, 2:163, 3:130, 4:81

role in locomotion, 3:90

Margulis, Lynn, 2:84

Marine animals

echolocation, 2:39–40

as entertainment, 2:7

exotic species, introduction, 2:91

external fertilization, 2:88

extinctions, 3:48

fossil record, 2:135

locomotion, 3:89–90

mammals, 3:90, 3:103

migration, 3:119

nitrogenous wastes, 2:89

reptiles, 4:112, 4:126

research, 3:104

tunicates, 4:130–131

zooplankton, 4:164

*See also* Marine biomes; *specific marine animals*

Marine animals, evolution

Cretaceous, 2:1

Devonian, 2:16–17

Jurassic, 3:59

Ordovician, 3:166–167

Permian, 4:2

Pliocene, 4:109

Silurian, 4:84

Triassic, 4:124–125

*See also* Tetrapods—from water to land

Marine biologist, 3:**104–105**

Marine biomes, 1:88–89

ecological relationships, 2:43

food web, 2:*127*

freshwater-saltwater transitions, 1:130–133, 3:57–58, 3:117–118

open ocean, productivity, 2:46

overuse by humans, 3:41–42

research, 3:104

seafloor biology, 1:95

seafloor spreading, 3:62

as thermophile habitat, 2:103, 2:127, 4:43, 4:126

and water pollution, 4:21

whale adaptations to, 3:136–139

*See also* Aquatic biomes; Tetrapods—from water to land

Marmots, foraging strategies, 2:131

Marmyridae (electric fishes), 3:168

Marshes. *See* Wetlands

Marsupials

Australia, 3:102, 3:163–164, 4:108

continental drift, 1:182, 1:183

crepuscular, 1:189

divergence from placentals, 4:8

extinctions, 3:48

fossils, 3:101

gliding and parachuting, 2:165

Ice Age migrations, 3:102, 4:15

as predators, 3:101

in South America, 4:109

Martins, as brood parasites, 3:55

Mastication (chewing)

inadequate, herbivores, 2:23

mammals, 3:141

role in digestion, 2:22

Mastodons, 4:15, 4:44

*Mastophora dizzydeani* (bolas spider), 2:130

Mathematical models. *See* Biomechanics; Biometry; Body size and scaling; Population models

Mating

assortative, 1:76

controlled, and selective breeding, 4:64–65

nonrandom, 1:76

repeated, and iteroparity, 3:58

sexual dimorphism and, 4:74–76

*See also* Copulation; Courtship; Sexual selection

Matrix

of antlers, 1:35

of blastula walls, 2:56

connective tissue, 1:129

keratin, 3:62

in mesenchymal cells, 3:109

of shells, 4:81

Mauthner's cells, 4:157

Max Planck Institute for Behavioral Physiology, 3:92

Maxillae, insects, 3:141

Maxillopoda. *See* Barnacles

Mayan societies

agriculture, 3:45

animals in folklore, 2:105

Mayr, Ernst, 3:**105–106**, 3:124

McDonald, J. H., 1:78

*The Meaning of Evolution*, 4:85

Medical doctor, 3:**107**

Mediterranean area, malaria, 3:94

Medulla oblongata, 3:157, 3:*157*

Medusa, 1:108, 1:156, 3:89, 3:111

Mefloquine, and malaria prevention, 3:97

*Megachasmidae* (megamouth sharks), 1:146

*Megachiroptera* (large bats), echolocation, 2:38

Megafauna, extinction, 4:44

*Megahippus*, as horse ancestor, 3:23

*Megaptera novaengliae* (humpback whale), migration, 3:119

Meiosis, 1:134–135

crossing over, 2:160–161, 4:46–47

defined, 2:53

gamete production, 2:88, 2:141–142

germ cells, 2:117

Melanin, in peppered moths, 3:185–187

Melatonin, 1:149–150, 2:66

Meliponinae (stingless bees), 1:37

Melon, in toothed whales, 2:39

*Melopsittacus undulatus* (budgerigars, budgies, parakeets), environmental impact, 2:94

Membrane potential, 3:159, 3:161

Menageries, 4:158

Mendel, Gregor, 1:75, 2:139, 2:159–161, 3:**107–109**

Mendelian genetics (Mendelism)

defined, 3:108

Modern Synthesis, 3:105–106, 3:123–125

*vs.* biogenic law, 4:144

Menstruation

and birth control pills, 2:67

and circadian rhythms, 1:151

and egg production, 2:117

hormone cycling, 3:18

and steroids, 2:67

Mercury pollution, 4:22

Meridional cleavage, 2:56

Merostomata (horseshoe crabs), 1:47, 1:48

Merozoites, role in malaria, 3:96

*Merychippus*, as horse ancestor, 3:23

Mesenchymal-epithelial transitions, 3:110

Mesenchyme, 3:**109–111**

Mesenchyme cells, 2:54, 2:55, 3:*110*

Mesenteries, defined, 1:102

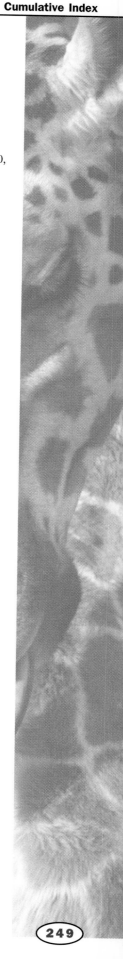

Osteoglossomorpha (bony-tongue fish), 3:168
Osteons, 1:111
Ostraciiform locomotion, 3:90
Ostracoderms, 4:134
Ostriches, pair bonding, 1:187
Other-feeders. *See* Heterotrophs
Otolithic membrane, 4:70
Ovaries, 2:65, 2:87, 3:19
Overchill Hypothesis, 4:17
Overharvesting
　and population decline, 2:61, 2:63
　regulations to prevent, 2:102
Overkill Hypothesis, 4:17
Oviduct, 2:50, 2:58
Oviparous reproduction
　Chondrichthyes, 1:144
　defined, 2:49
　monotremes, 3:101
　reptiles, 4:51
*Ovis*
　*canadensis nelsoni* (desert bighorn sheep), 4:148
　*orientalis* (Asiatic moufflon), 2:28
Ovoviparous reproduction, 1:144, 3:113
Ovulation, and birth control pills, 2:67
Ovum. *See* Egg
Owen, Richard, 3:15
Owls
　acoustic signals, mimicry, 2:15
　interspecies communication, 1:5
　niche differentiation, 4:74
　as nocturnal, 3:162
　northern spotted, population decline, 2:62, 2:64, 3:25–26
　pair bonding, 1:186
Oxen, horns, 1:35
Oxidative phosphorylation, 1:136
Oxygen
　absorption, 1:2, 2:163–164, 4:11, 4:52, 4:53
　anaerobes, 4:22, 4:42–43
　in blood, 1:97–99
　generated by plants, 1:178, 2:48
　limited, extremophile tolerance, 2:103
　marine ecosystems, contribution, 1:88
　and phytoplankton blooms, 2:48
　requirement for, 4:55–56
Oxyhemoglobin, 1:98
Oxytocin, function, 3:18
Oysters
　commensal relationships, 3:56

Cretaceous, 2:1
　as filter feeders, 2:115
Ozone, as greenhouse gas, 2:169, 4:19
Ozone layer, depletion, 2:74, 4:164–165

# P

p53 gene, 1:124
P'a Ku, mythological character, 2:105
Pack rats (wood rats), water economy, 4:146, 4:148
Pack species, altruistic behavior, 2:119
Paddlefish, 3:168
Paedomorphosis, 3:8–10, 4:131
Pain, felt by animals, 1:25–26
Paine, Robert, 3:64–66
Pair bonding
　cooperative mating, 1:187–188
　lekking, 1:187–188
　monogamy, 1:186–187, 4:79
　polyandry, 1:187, 4:75
　polygamy, 1:187, 4:79
　polygyny, 1:187, 4:77
　promiscuity, 1:187–188
Palearctic region, continental drift, 1:73–74
Paleoanthropology, 3:32
Paleobotany, 3:172
Paleocene epoch, 2:162, 3:138, 4:107, 4:108
Paleoecology, 2:43, 3:172, 3:174
Paleontologist, 3:**172–174**
　errors in determining extinctions, 2:101
　George Gaylord Simpson, 4:85–86
　identifying new species, 2:98
Paleontology, 3:**174–175**, 3:184
Paleozoic era, 2:162, 4:124
Pancreas
　blood sugar control, 2:66, 3:19
　role in digestion, 2:19, 2:20, 2:22
Pancreatic amylase, 2:19
Pancreatic islet cells, 2:65
Pandas, 4:113, 4:152, 4:160
Pangaea, 1:181
　Carboniferous, 1:127
　Cretaceous, 2:1
　Jurassic, 3:59, 3:101
　Paleozoic, 4:124
　Permian, 4:1–2, 4:108
*Panthera leo spelaea*, 4:15
Panthers, as threatened, 4:113
Paracanthopterygii, 3:170–171

Parachuting, 2:165–166
*Parahippus*, as horse ancestor, 3:23
Parakeets, environmental impact, 2:94
Parallelism (apomorphic traits), 1:184
Parapodia, 1:31
Parasites, 3:54–55, 3:**175–180**
　antagonistic relationships, 1:169
　brood, 3:51, 3:53, 3:176–178, 3:179
　colonization, 1:162
　ectoparasites, 1:32, 1:38, 1:47, 1:48, 3:175–176
　endoparasites, 1:32, 3:175–176
　energy use, 1:96
　evolution and, 3:178–179, 4:137
　facultative, 3:176
　of honeybees, 1:37–38
　of insect pests, 1:78–80
　interactive relationships, 1:169, 4:36
　intermediate hosts, 3:176–177
　keystone effect, 3:66
　long-term commensals as, 3:56
　macroparasites, 3:176
　malacostraceans, 1:49
　microparasites, 3:176
　monoxenous life cycles, 3:176
　nematodes as, 3:152
　obligatory, 1:141–142, 3:176, 4:43
　phylogenic studies, 3:129
　*Plasmodium* as, 3:94–96, 3:95, 3:177
　population ecology, 4:30
　Red Queen Principle, 1:160
　role in mating behaviors, 3:179–180
　social, 3:178
　tapeworms as, 3:54–55, 3:176, 3:177
　trematodes, 4:13–14, 4:122
　vectors, 3:177
　vertebrates as, 3:176, 3:177
　*See also* Viruses
Parasitoids, 3:175, 4:36
Parasitology, defined, 3:104
Parasympathetic divisions, 3:156
Parathyroid glands, 2:65, 2:66
Parazoa, body plan, 1:106
Parental imprinting, defined, 3:51
Parent-young interactions
　altruism, 4:96–97
　brood parasites, 3:51, 3:53, 3:177–178, 3:179
　captive condors, 4:30

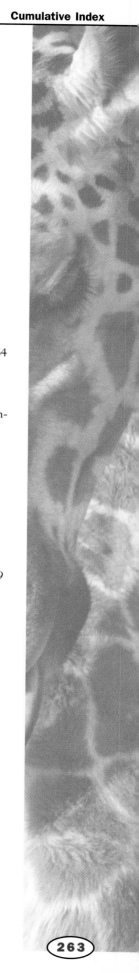

Ruminants
    defined, 2:28
    endosymbiotic fauna, 2:69–70,
        4:128
    fossil record, 4:109
    herbivores as, 2:23, 2:28–29,
        2:69, 2:110, 2:113, 4:128
Runaway selection model, 4:77–78
Running locomotion, 3:86, 3:88,
    3:102
Russell, William, 1:28
Russia, malaria, 3:94

## S

Saber-toothed cats (*Smilodon*), 3:102,
    4:15, 4:34, 4:109
Sacropterygians, 3:80
SAD (seasonal affective disorder),
    1:150
Safe Drinking Water Act, 4:24
Sagital plane, defined, 4:133
Salamanders
    axolotls, 1:21, 3:9, 3:10
    chemical deterrents, 1:39, 2:15
    defined, 1:19
    gills, 2:163
    heterochrony, 3:9
    locomotion, 3:90
    metamorphosis, 3:10, 3:113
    paedomorphosis, 3:9
    phylogenetic trees, 4:7
    reproductive strategies, 2:49
    respiration, 4:54
    tiger, as axolotl ancestor, 3:10
    *See also* Amphibia
Salinization
    of soils, 2:75–76
    of water supply, 4:100
Saliva
    digestive enzymes, 3:140
    function, 2:19
    of insects, 3:141
Salivary glands, role in digestion,
    2:22
Salk, Dr. Jonas, 1:26
Salmon
    effect of acid rain, 2:81
    effect of DDT, 2:12–13
    fish ranching, 1:44
    habitat restoration, 3:3–4
    migration, 1:130–133, 3:117,
        3:120, 3:170
    as semelparous, 3:57–58, 3:58
    *See also* Catadromous—diadro-
        mous and anadromous fishes
*Salmonella*, 3:67

Salmonids (Salmoniformes), 3:170
Salt concentration, of urine, 4:147
Salt lakes, extremophile tolerance,
    2:103
Salt-secretion cells, diadromous
    fishes, 1:131
Saltwater-freshwater transitions,
    1:130–132, 3:57–58, 3:117–118
San Andreas Fault, 1:182
San people, as hunter-gatherers,
    3:43, 3:44
*A Sand County Almanac*, 1:67, 3:4
Sand dollars, 2:36
Sandpipers, pair bonding, 1:187
Sarcomeres, 3:142, 3:144
Sarcopterygii. *See* Lobe-finned fishes
Sargasso Sea, eel migration,
    3:117–118
Sargent, Theodore D., 1:158
Sauria. *See* Lizards
Saurischians, 2:24
Sauropods, 2:2, 2:24, 3:60
Savanna, 1:90–91
Sawfish (*Pristidiformes*), 1:146
Scabies mite, 3:31
Scale, cottony cushion, 1:81
Scales
    bird legs, 4:62
    ctenoid, 4:61
    evolution, 4:112
    ganoid, 4:61
    mesodermal origin, 4:61
    placoid, 1:144, 4:61
Scales, feathers, and hair, 4:**60–63**
    *See also* Feathers; Hair and fur
Scallops (pectens)
    eyes, 3:133
    movement, 3:82–83, 3:89
Scaphopoda (tooth shells), 3:130,
    3:134
Scaridae (parrotfishes), 3:171
Scavengers
    crustaceans, 1:48
    defined, 2:6
    dogs as, 2:5, 2:28
    habitat and, 2:177, 3:163
    maritime, role in food web, 2:127
    omnivores as, 2:111
    Turbellaria, 4:13
Scent glands, 1:165, 3:100
Schistosomiasis, 3:177
Schizocoelous development, 1:102,
    1:103, 2:54
Schooling, of fish, 1:121
*Schreckstoff* (fright substance), 3:170
Schwann cells, function, 2:55, 3:153
Sciaenidae (drum fishes), 3:171

Scientific illustrator, 4:**63–64**
Scientific method
    and comparative biology,
        1:170–171
    and peppered moth research,
        3:187–188
Scientific names, 1:64
Sclera, 4:139
Scleroblasts, 3:110
Scleroglossa, 4:51
Scolex, of cestodes, 1:141, 1:142
Scombridae (tuna and mackerel),
    3:172
Scopelomorpha (lanternfishes),
    3:172
Scorpaeniformes (sculpins, rockfish),
    3:171
Scorpions
    characteristics, 1:48
    classification, 1:47
    Devonian, 2:17
    as nocturnal, 3:162
    Silurian, 4:84
    success of, 2:26
*Scr*, and hox genes, 4:72
Sculpins (Scorpaeniformes), 3:171
Scyphozoa. *See* Jellyfish
*The Sea Around Us*, 4:82
Sea cows, evolution, 3:164
Sea cucumbers. *See* Echinodermata
Sea hare, internal shell, 4:80
Sea levels, rising and dropping,
    2:167–169, 4:44, 4:124
Sea lilies, 2:36–37, 3:167, 4:84, 4:125
    *See also* Echinodermata
Sea otters
    as keystone species, 3:65
    tool use, 4:114
Sea scorpions
    Devonian, 2:17
    Ordovician, 3:166
    Silurian, 4:84
Sea squirts (Branchiostomates),
    1:129, 4:130
    *See also* Urochordata
Sea urchins, 2:36, 2:56–57
    *See also* Echinodermata
Seafloor biology, 1:95
Seafloor spreading, role in extinc-
    tions, 3:62
Seagulls, behavior, 1:17–18
Seahorses (Syngnathidae), 3:171
Seals
    dominance conflicts, 2:31
    evolution, 3:164
    locomotion, 3:90–91
    sexual dimorphism, 4:74

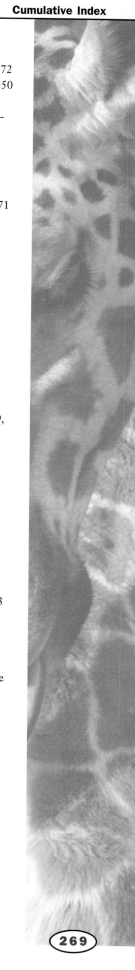

Vultures
phylogenetic tree, 4:*6*
soaring, 3:89
tool use, 4:114, 4:115
wing shape, 2:124

# W

Walcott, Charles, 1:112, 1:114, 1:116
Walking locomotion, 3:85–86
Wallace, Alfred Russel, 1:58, 2:10, 3:98–99, 4:**145–146**
Wallace's Line, 4:145
Walruses, social interactions, 4:89
War of attrition strategy, 1:12
Warfarin, resistance to, 4:5
Warm-blooded (homeothermic) animals
advantages, 3:163
birds, 1:52
defined, 3:100
global cooling and, 4:109
mammals, 1:52, 3:100
tuna and mackerel, 3:172
Warning calls, 1:3–4, 1:121, 2:15
Warning coloration. *See* Aposematism; Batesian mimicry; Müllerian mimicry; Visual signals
Wasps
genetic similarities, 4:96
homing behavior, 2:83
mimicry of, 1:161, 1:164
parasitic, 1:169
parthenogenesis, 2:88
tool use, 4:114
*See also* Apidae
Waste products
in blood, 1:97
of desert animals, 4:147, 4:149
of diadromous fishes, 1:131
in eggs, 2:49–50, 2:53, 2:89
excretion, trematodes, 4:123
feces, 2:21, 2:22–23, 2:113, 2:129–130
fossilized (coprolites), 2:135
olfactory signals, 2:129–130
urea, 2:89
uric acid, 2:53, 2:89
*See also* Excretory and reproductive systems
Waste products, man-made
agricultural, 4:100, 4:101
domestic sewage, 1:69, 2:73, 4:20–22
industrial, 4:21–23
and water pollution, 3:148, 4:20–23

Water
chemically synthesized by wood rats, 4:146, 4:148
competition for, 1:173, 3:42
cycling in ecosystem, 2:47
diversion, to reservoirs, 2:48, 2:75
drainage, in restored habitats, 3:3
as habitat requirement, 3:12
percentage of Earth covered, 1:87
quality, conflicts, 4:20
quality, in restored habitats, 3:1–4
as renewable resource, 3:148
sound, movement through, 1:165, 2:39–40
*See also* Aquatic biomes; Marine biomes
Water balance. *See* Moisture balance; Osmoregulatory systems
Water buffalo, domestication, 2:29
Water development projects, and habitat loss, 2:182, 2:184, 4:113
Water economy in desert organisms, 4:**146–149**
adaptations, animals, 4:149
adaptations, plants, 4:146–147
canyon mouse, 4:148
crepuscular species, 1:189–190
desert bighorn sheep, 4:148
frogs, 1:22, 3:9
kangaroo rats, 1:6–7, 4:147–149
killifishes, 3:170–171
kit foxes, 4:146
limiting resource, 1:170
phainopepla, 4:148
roadrunners, 4:148
wood rats, 4:146, 4:148
Water fleas (Branchiopoda), 1:48
Water pollution
causes, 3:148, 4:20
classifying, 4:20
environmental impact, 1:68–69, 2:44, 2:*78*
hydroelectric plants and, 3:147–148
impact on plankton, 4:13
oil spills, 2:73, 4:23, 4:33, 4:113
organic waste, 4:22
and population distribution, 2:76
prevention, 4:100–101
radioactive materials, 4:22–23
and reduced biodiversity, 3:25
regulations to control, 4:20
Safe Drinking Water Act, 4:24
synthetic chemicals, 4:22

thermal, 4:23
treatment methods, 4:20–21
*See also* Fertilizers, chemical; Pesticides
Water solubility
of hormones, 3:17
of keratin, 3:63
Water supply
damaged by exotic species, 2:93, 4:20
drinking water, 4:20, 4:24, 4:100
shortages, 2:79–80, 3:42, 4:33, 4:100
Water vapor, as greenhouse gas, 2:169
Waterfowl
damaged by oil spills, 4:*21*, 4:23
migratory, declining populations, 2:44, 4:23
Watson, James, 3:5
Wavelength (sound), and echolocation, 2:38
Wax, produced by bees, 1:37
Weapons, hunting, 3:46–47
Weberian ossicles, 3:170
Webs, of spiders, 1:47–48, 1:63
Wegener, Alfred, 1:181
*Weigeitosaurus* (flying lizard), 2:*166*
Weinberg, Wilhelm, 1:75
Western blot technique, 3:127
Wetlands
characteristics, 1:88
to clean waste runoff, 1:67
draining, 2:44, 2:72, 2:180–181, 3:1, 3:25, 4:113
fossil preservation in, 2:136
productivity, 2:46
restoring, 3:3
Whales
Archaeoceti, 3:137
baleen, 2:38, 2:115, 3:103, 3:136–138, 4:164
barnacle hitchhikers, 1:174–175
blue, 2:113, 2:115, 3:103, 3:119
echolocation, 2:38, 2:39–40
as entertainment, 2:7
hair, 3:100, 3:138
human interference, 1:6
humpback, 3:*137*
hunting for, 3:47, 3:103
locomotion, 3:90
migration, 3:119
morphological evolution, 3:**136–139**, 4:109
Mysticeti, 3:136–137, 3:138
and noise pollution, 4:23
Odontoceti, 3:136, 3:138